North American Mixing Forum 2013

Core Programming Area at the 2013 AIChE Annual Meeting: Global Challenges for Engineering a Sustainable Future

San Francisco, California, USA
3 – 8 November 2013

ISBN: 978-1-63439-048-4

Printed from e-media with permission by:

Curran Associates, Inc.
57 Morehouse Lane
Red Hook, NY 12571

Some format issues inherent in the e-media version may also appear in this print version.

Printed by Curran Associates, Inc. (2014)

AIChE
3 Park Avenue
New York, NY 10016-5991

Phone: (203) 702-7660
Fax: (203) 775-5177

www.aiche.org

Additional copies of this publication are available from:

Curran Associates, Inc.
57 Morehouse Lane
Red Hook, NY 12571 USA
Phone: 845-758-0400
Fax: 845-758-2634
Email: curran@proceedings.com
Web: www.proceedings.com

TABLE OF CONTENTS

(70a) Novel High Efficiency Axial Flow Impeller 1
Robert W. Higbee, Jason J. Giacomelli, Richard K. Grenville

(70b) Concentration and Mixing Effects On the Production of Amine Hydrochloride Salts in a Confined Impinging Jet Reactor 2
Navid Ershad, Suzanne Kresta

(70c) Characteristics of Fluid Deformation Induced By a Rotationally Reciprocating Impeller 3
Saki Senda, Yoshiyuki Komoda, Hiroshi Takeda, Yushi Hirata, Ruri Hidema, Hiroshi Suzuki

(70d) The Netmix Reactor: Heat and Mass Transfer Modeling and Mixing Properties Assessment 6
Carlos M. Fonte, Marcelo F. Costa, Madalena M. Dias, José Carlos B. Lopes

(70e) The Simplest Stirred Tank for Laminar Mixing: Mixing in a Vessel Agitated By An Off-Centered and Angled Disc 9
Mario M. Alvarez, David Bulnes-Abundis

(70f) Low Power, Baffle-Free Mixing With Contra-Rotating Impellers 10
Justin Register, John A. Regalbuto, John R. Regalbuto

(70g) New Impeller Type for Significant Mixing Efficiency Improvement and Energy Saving 11
Li Wang

(132a) An Unusual State During the Production of Solid Stabilized Emulsions in Stirred Tanks: Paste 12
Inci Ayranci, Louis Fradette

(132b) Macro-Mixing of the Dispersed Phase in An Immiscible Liquid-Liquid Stirred Reactor 13
Chao Yang

(132c) The Effect of Particle Shape On the Scale Up of Just Suspended Impeller Speed 14
Matthew Dimattesa, Daniel Cendo, Steven Boesch, Arthur W. Etchells, Robert P. Hesketh

(132d) Effect of Impeller Submergence, Position, Diameter and Speed On Flow Regime, Surface Air Entrainment and Mixing Effectiveness of Partially Filled Stirred Vessels 15
Shilan Motamedvaziri, Piero M. Armenante

(132e) Drop Breakage By Shear Tip Speed 17
Thomas L. Rodgers, Michael Cooke

(132f) A Process-Based Analysis On the Generation of Pickering Emulsions 18
Emir Tsabet, Louis Fradette

(132g) The Taylor Dispersivity of a Passive Tracer in the Pressure-Driven Flow of a Concentrated Suspension of Rigid, Non-Colloidal Spheres 19
Arun Ramachandran

(132h) Improving Euler-Granular Models in Multiphase Mixing for Polydispersed Flows 21
Brigette Rosendall

(188a) Perspectives On the Use of Simulation and Correlation for Design of Mixed Chemical Reactors 22
Gary K. Patterson

(248a) Synthesis of Fine Chemicals by Heterogeneous Catalysis: A Critical Approach of the Role of Solvent for Hydrogenation Reactions 23
Carlos Apesteguía

(259a) Laminar Flow Pattern and Heat Transfer in a Vessel With Eccentrically Positioned Multiple Inclined Blade Impellers 24
Mathijs Koot, Harry E. A. Van Den Akker

(259b) The Effects of a Spatially Variant Velocity Field On Stretching: Intuitive Measures 25
David Bigio, Jason Nixon

(259c) Measuring Blend Time In Laboratory Agitated Vessels: A Case for the Acid-Base-Indicator Method 26
Julian B. Fasano, Kevin J. Myers, Eric E. Janz

(259d) The Confined Impeller Stirred Tank: Blend Time 27
Marcio Bezerra Machado, Suzanne Kresta

(259e) Mixing Time in Partially Filled Stirred Vessels 28
Shilan Motamedvaziri, Piero M. Armenante

(259f) Impeller Placement Optimization; Mixing Versus Mechanical Shaft Fatigue 29
Sang Jin Lee, Robert Higbee, Binxin Wu

(259g) Simulations of Mobilization of Bingham Layers in Agitated Tanks 30
Jos Derksen

(316a) Formation of Poly(ε-caprolactone) Nanoparticles in Microstructured Mixers Via Solvent Replacement Precipitation: Fundamental Studies and Scale-Up Concepts 31
Andreas Kölbl, Abdul Qayyum, Manfred Kraut, Roland Dittmeyer
(316b) Bioreactor Simulation With Cuda 32
Christian Witz, Tawan Tantikul, Johannes G. Khinast
(316c) Characterization of Kla Using Definitive Screening Designs for Fermentation Development and Scale-Up 34
Dogan Ornek, Rayford Anderson, Michelle Chopra, Wenglong Lin, Philip Ramsey
(316d) Scaling Down the Production of Biopharmaceuticals to Continuous Micro-Chips 35
Mario M. Alvarez, Leydi Maribel Carrillo-Cocom, Lucía D. Garza-García, Eduardo J. Tapia-Mejía, Sergio Camacho-León, Ciro Angel Rodríguez-González, Erika García-López
(316e) Computing Dissolved Oxygen Profiles in Aerobic Fermenters 36
Gopal Kasat, Damodaran Vedapuri, Gregory Benz
(316f) Hydrodynamic Effects of An Arch-Shaped Fiber Optic Probe in a USP Dissolution Testing Apparatus 2 37
Yiran Zhang, Gerard Bredael, Piero M. Armenante
(371a) Towards the Combination of Modeling Strategies for Solid-Liquid Mixing 38
Bruno Blais, Francois Bertrand
(371b) CFD Modeling of Viscoelastic Droplet Breakup 40
Laura J. Dietsche, Nilesh Parmar
(371c) Hybrid Multiphase CFD Simulation for Interfacial Area Prediction in Liquid-Liquid Extraction 41
Kent E. Wardle
(371d) Determination of Drop Size Distribution in Turbulently Agitated Liquid-Liquid Dispersion Using Numerical Simulations 42
Alexandra E. Komrakova, Dmitry Eskin, Jos Derksen
(371e) CFD Simulation of Integrated Physical and Biological Processes for Anaerobic Digestion 43
Binxin Wu
(371f) Application of a Combination of Statistical Techniques and Mechanistic Modeling to the Mixing Process in Co-Rotating Twin Screw Extruders 44
Osaighe Kennedy Amedu
(371g) Computational Study of Solids Mixing in Flotation Cells Using CFD-DEM Simulations 45
Manjunath Basavarajappa, Sanja Miskovic
(371h) Erosion of Granular Beds - Solids Suspension Mechanisms 46
Jos Derksen
(467a) Experimental Investigation of Multiphase Mixing in Annular Centrifugal Contactors Using High-Speed Imaging and Electrical Resistance Tomography (ERT) 47
Kent E. Wardle
(467b) Mixing Performance of a Bioreactor 48
Sally Mattar, Marcio Bezerra Machado, Suzanne Kresta, Dominic Sauvageau
(467c) Gas Dispersion With Up-Pumping Maxflo W Impellers 49
Muneeb Ur Rahman Shaik, Kevin J. Myers, Eric E. Janz
(467d) The Simulation of Solid Suspension By Using Scaba and A310 50
Ruey Chi Hsu, Jer Chang Kang
(467e) Improved Mixing and Mass Transfer Scale-Up: Design & Characterization for Cell Culture in Single Use Bioreactors 51
Nephi Jones, Christopher Brau
(467f) Foamless and Bubble Free Dynamic Gas-Liquid Mixing in Cell Culture Bioreactor 52
Alan Cheng, Ying Zhou
(467g) Vortex Generation With Submerged Baffles: Comparison Of Two Scales 53
Jason J. Giacomelli, Wojciech Wyczalkowski
(476a) Computational Fluid Dynamic Simulations of Gas-Liquid and Liquid-Liquid Flow in An Advanced-Flow Reactor (AFR) 54
Maria Jose Nieves-Remacha
(476b) Uncertainty Quantification in Coal Gasifier Simulations For Reliable Predictive Computational Fluid Dynamics Models 57
Aytekin Gel, Mehrdad Shahnam, Arun K. Subramaniyan
(476c) Experimental and CFD Study of Methane Steam Reforming in a Catalytic Plate Reactor and Characterization of the Coated Noble Metal Catalyst 59
Mayur Mundhwa, Brant. A. Peppley, Christopher P. Thurgood

(476d) Three Dimensional CFD-Based Numerical Study of Chemically Reacting Carbon Particles Within a Randomly Packed Bed 60
Sebastian Schulze, Petr A, Nikrityuk, Bernd Meyer
(476e) Development of a Methodology to Obtain the Kinetic Mechanism of Steam Methane Reforming Compact Reactor Using CFD Coupled With Hybrid Genetic/Surface Response Optimization Model 61
Rodrigo P. M. Moreira, Lucilla Almeida, Raphael David A. Bacchi, Karolline Ropelato, Fabio M. Passarelli, Erick F. A. Quintella Coelho, Antonio Marcos F. Bidart
(476f) Numerical Investigation of Packed Bed Reactors With Non-Spherical Particles 62
Thomas Eppinger, Nico Jurtz, Matthias Kraume
(476g) Computational Fluid Dynamics for the Analysis and Optimization of a Lab-Scale Stirred Tank Reactor With Catalytic Basket 64
Cláudio P. Fonte, Maria Braga, Léna Brunet-Errard, Isabelle Pitault, Serge Simoëns, Claude De Bellefon, Vania Santos-Moreau, José Carlos B. Lopes
(578a) CFD Model of Suspension Polymerization in Mixed Reactors 65
Michal Vonka, Miroslav Soos, Massimo Morbidelli
(578b) Simulation of Mixing in Hot Melt Extruders Using Smoothed Particle Hydrodynamics 66
Andreas Eitzlmayr, Stefan Radl, Gerold Koscher, Johannes G. Khinast
(578c) Investigation of Single and Multiphase Fluid Flow Behavior in Stirred Tanks By Means of CFD and Radioactive Particle Tracking 68
Hamed Bashiri, Ebrahim Alizadeh, François Bertrand, Jamal Chaouki, Mourad Heniche
(578d) Computational Fluid Dynamics Investigation of Mechanical Mixing in Anaerobic Bioreactors 69
Jia-Jun Wang, Chun-Yan Ge, Xue-Ping Gu, Lian-Fang Feng, Binxin Wu
(578e) Validation of Openfoam Solver for Mixing Tank Analysis 70
Gopal Kasat, Damodaran Vedapuri, Pritam Hule, Yogesh Bapat
(578f) Steady State CFD for Determining Just Suspended Speed 71
David C. Russ, R. Eric Berson
(578g) Discrete Tangent Based Sensitivity Analysis Method in Computational Fluid Dynamics (CFD) for Stirred Tank Reactors 72
Vibha Srinivasan, Prabhu Sathyamurthy, Sanjay Mathur
Author Index

(70a) Novel High Efficiency Axial Flow Impeller

Higbee, R. W., Philadelphia Mixing Solutions Ltd.
Giacomelli, J. J., Philadelphia Mixing Solutions Ltd.
Grenville, R. K., Philadelphia Mixing Solutions Ltd.
Originally presented on: 11/4/2013 8:30:00 - 8:50:00

For the class of axial pumping impellers, where the design goal is to maximize the amount of input energy going to pumping, as the certainty of the nature of the impeller inlet velocity profile and characteristics of the desired exiting velocity profile increase, the magnitude of the hydraulic energy efficiency will increase when the impeller designer has the ability to create unique blade pitch face cambers at all radial blade locations. A specific example of this design philosophy is presented for the case of an impeller in a motionless fluid reservoir where the goal is to axially pump a cylindrical column of fluid whose velocity profile is completely uniform with no rotational velocity component. A theoretical maximum attainable power number versus pumping number efficiency curve for this class of axial flow impellers will be plotted and will be shown to be linear on a log-log scale and will be compared to a previously defined British Hydrodynamic Research Group (BHRG) curve which also revealed a logarithmic linear relationship between power number and pumping number for many commercially available axial impellers. The zone between the BHRG line and the higher proposed maximum efficiency line will be defined as the design space within which any new "high efficiency" impeller must reside. A family of new axial flow impellers, the patented PMSL APII series, will be presented whose power and pumping numbers form a curve that will be shown to reside in this currently unoccupied high efficiency design space. The impeller's pumping capacity was measured using 2-D Particle Image Velocimetry. Impeller power draw was measured using a stationary torque sensing load cell and tachometer.

(70b) Concentration and Mixing Effects On the Production of Amine Hydrochloride Salts in a Confined Impinging Jet Reactor

Ershad, N.,
Kresta, S., University of Alberta
Originally presented on: 11/4/2013 8:50:00 - 9:10:00

A Confined Impinging Jet Reactor (CIJR) was used to react 4, 4'-methylene dianiline (MDA) with hydrogen chloride (HCl) to determine whether methylene dianiline dihydrochloride (MDA.2HCl) and/or methylene dianiline monohydrochloride (MDA.HCl) form under various reaction conditions. The two aminehydrochloride (AHC) salts are byproducts of a desired reaction sequence in the manufacture of polyurethanes. The effects of several processing variables (local concentration of MDA and HCl) and the intensity of local mixing on the formation of AHC salts were studied. In almost all cases, solid precipitates formed. They were characterized using a wide range of analytical techniques: elemental analysis (EA), FTIR, TGA, DSC, Optical Microscopy, SEM and EDX. The most successful analytical methods and the effects of each process variable on the solids composition, thermal properties, morphology and average size are discussed. Results show that formation of AHC salts more than anything depends on local concentration of MDA and it dominates the structure of product in terms of formation of MDA.HCl or MDA.2HCl. Mixing intensity on the other hand, affects purity of product and evenness in shapes and average size of particles.

(70c) Characteristics of Fluid Deformation Induced By a Rotationally Reciprocating Impeller

Senda, S., Kobe University
Komoda, Y., Kobe University
Takeda, H., Rflow Co., Ltd.
Hirata, Y., Osaka University
Hidema, R., Kobe University
Suzuki, H., Kobe University
Originally presented on: 11/4/2013 9:10:00 - 9:30:00

Although poorly mixed regions or stagnant regions in a mixing vessel are avoidable in turbulent mixing condition due to velocity field fluctuation, effective mixing performance is achieved by an intentional change in velocity field or unsteady flow field even at a laminar mixing condition. For example, some researchers already have reported that the mixing operation with periodic changes in rotational speed or rotational direction improved mixing performance. However, most of them have employed the impeller motion with comparatively high rotational speed and long period of operational condition change. Therefore, unsteady velocity field appeared only in the moment around the operational condition change and the unsteady impeller operation will not play significant role on the enhancement of overall mixing is relatively small. On the contrary, the impeller motion with continuous change in rotational speed from negative to positive has a possibility to make the most of unsteady velocity field for improving mixing performance. In this study, this mixing operation was named “rotationally reciprocating mixing” after the impeller motion with rotating back and forth. Since the mixing process with unsteady velocity field is usually three dimensional and complicated, it is difficult to describe the mixing mechanism induced by the unsteady velocity field. Therefore, we have firstly chosen two-dimensional mixing process in r-$q$plane of a cylindrical vessel in order to neglect the effect of vertical fluid motion on the mixing mechanism. Since Inoue reported that a streak line should be a template of laminar mixing process (Inoue et al, 1999,2000), streak line was numerically and experimentally visualized in this study to elucidate the mixing process.

In our previous study, we have already experimentally and numerically clarified that the rotationally reciprocating plate impeller generates several pairs of vortices in the r-$q$plane of the cylindrical vessel and realizes an excellent fluid deformation process especially at the reciprocating condition of large amplitude and long period (Komoda et al, 2012). These vortices alternately generate suction and discharge flows in radial direction. A fluid lump was dragged inward due to the suction flows and then pushed outward due to the discharge flow. Therefore, a fluid lump is effectively transported in radial direction in spite of rotational impeller motion. Additionally, a fluid lump discharged from the impeller tip was effectively stretched behind the impeller plate and then folded due to the inverse motion of fluid flow after the impeller counterturn. The mixing process was clearly observed by the development of streak line. As a result, we found that the vortex generated plays a significant role on the deformation and transportation of fluid in a mixing vessel.

In the case of a typical unidirectional rotational impeller, various characteristics in the mixing process are characterized by mixing Reynolds number. At the condition of the same dimensions of mixing equipment with the same test fluid, the rotational speed becomes the only one parameter which affects the mixing process. Similarly, in the rotationally reciprocating impeller, the mixing process should be drastically changed with reciprocating period. At least, the flow regime must be changed from laminar to turbulent with the decrease in reciprocating period, and the mixing mechanism must change as well. Therefore, in the present study, we have investigated the two-dimensional fluid deformation process in a cylindrical vessel induced by a plate impeller rotationally reciprocating at different period from the view point of experimental visualization and numerical simulation.

Firstly, we have experimentally visualized the development of streak line. A flat-bottomed cylindrical vessel made of transparent acrylic resin having the inner diameter of 0.08 m was filled with an aqueous starch solution with the liquid height of 0.07 m. Flat plate impeller having the width of 0.06 m and the height of 0.08 m was rotated back and forth at the center of the vessel just above the bottom. Since the impeller height is larger than the liquid height, the velocity field can be considered as 2-dimensional in the r-θ plane far above the vessel bottom. The impeller motion was preciously controlled by a stepping motor via a mixing shaft (o.d. = 8 mm). The orientation of the impeller is expressed by $j = A\ [1\text{-}\cos(2\pi t/T)]$, where A and T represent the amplitude and period of rotational reciprocation of the impeller. In this study, the reciprocating amplitude was constant at $\pi/2$ where the impeller sweeping region corresponds to a circle, and the reciprocating periods were 0.5, 1, 2, 4s. The density and viscosity of the aqueous starch solution used were 1220 kg/m^3 and 0.1 Pa·s, respectively. After attaining periodically stable velocity field by reciprocating the impeller for several cycles, we have started injecting the tracer fluid, which is the starch solution containing a fluorescent dye (Rhodamine B), into the starch solution in the mixing vessel at the point 1 cm below the liquid surface and 8 mm from the vessel wall. By using a planar laser light source (532nm), the track of the tracer fluid in the r-θcross-section was visualized. The streak line thus obtained was observed from the vessel bottom via a mirror.

We also have numerically carried out the visualization of streak line under the same reciprocating condition with experiments. Periodically stable sets of 2-dimensinal velocity fields in the r-$q$plane have been obtained for each reciprocating condition under laminar flow regime using CFD software RFLOW. Tracer particles put at the tracer injection point one after another for each calculation time step, and then they were moved according to the corresponding velocity field. In the course of the calculation, another tracer particles were inserted at the middle point of two tracer particles in case of too large particle separation in order to obtain smoothly deformed streak lines, which can be drawn by connecting all adjacent tracer particles.

In the velocity profile obtained in CFD simulation, the generation of vortices behind the impeller tip just after the impeller counterturn still has been observed for each reciprocating period investigated. However, the development process of the vortices significantly differed with reciprocating condition. At the short reciprocating periods of T = 0.5 and 1s, the shape of vortices deformed in circumferential direction, their center were released from the impeller tip, and they remained at the next impeller counterturn. At the period longer than 2s, the vortices have already disappeared at the next impeller counterturn and their center always stayed at the impeller tip at the longest period condition.

After the start of tracer fluid injection at t= 0, the development of the streak line was observed by means of experimental and numerical methods. As already reported, the tracer fluid was dragged inward and then strethced along the vessel wall at the case of T = 2s. As a result, the corss sectional area of the vessel was almost filled with the deformed streak line because the streak line was effectively stretched and folded due to developed vortices. However, the central region near the mixing shaft remained without the instrusion of streak line because the fluid lump near the shaft stretched along the impeller plate but eventually pushed outward from the impeller tip after many cycles of reciprocation. At a shorter period of T = 1s, the deformation of the streak line proceed rapidly but was similar to that of T = 2s with the generation of poor mixing regions. On the other hand, at the longest period of T = 4s, the streak line was mainly stretched and folded along the vessel wall around the impeller tip because vortices were not separate from the impeller tip. The characteristic behavior of the deformed streak line has been successfully expressed in the numerical simulation at the period longer than 1s. However, at the shortest period of T = 0.5s, the definitely different deformed streak lines were obtained in numerical and experimental observations. The numerical streak line was still similar to that at T = 1s and 2s, and poor mixing regions remained near the shaft. In the experiment, however, the streak line was initially stretched and folded in the similar manner with numerical simulation but was observed to be split into several deformed streak lines. As the small deformed streak lines spread all over the vessel, poor mixing regions were disappeared. Additionally, a regular deformation of stretching and folding of streak line have not been performed anymore. That is to say, the velocity field contained terrible fluctuation and showed three-dimensionality. From these experimental results, turbulent flow was generated in the mixing vessel at this condition and the velocity field obtained in CFD simulation under a laminar flow condition is not appropriate.

At the condition of equipment dimensions and fluid properties investigated in this study, a significant enhancement in fluid deformation and an elimination of poor mixing regions have not been achieved as far as in a laminar flow condition. Additionally, at the longest reciprocating period, the fluid deformation was localized around the impeller tip. On the other hand, under a turbulent flow regime at the shortest reciprocating period, poor mixing regions immediately disappeared.

(70d) The Netmix Reactor: Heat and Mass Transfer Modeling and Mixing Properties Assessment

Fonte, C. M., Faculdade de Engenharia da Universidade do Porto
Costa, M. F., LSRE - Laboratory of Separation and Reaction Engineering - Faculty of Engineering - University of Porto
Dias, M. M., LSRE - Laboratory of Separation and Reaction Engineering - Faculty of Engineering - University of Porto
Lopes, J. C. B., LSRE - Laboratory of Separation and Reaction Engineering - Faculty of Engineering - University of Porto
Originally presented on: 11/4/2013 9:30:00 - 9:50:00

The NETmix Reactor: Heat and Mass Transfer Modeling and Mixing Properties Assessment

C. M. Fonte, M. F. Costa, M. M. Dias, J. C. B. Lopes

Laboratory of Separation and Reaction Engineering, Faculty of Engineering, University of Porto, Porto, Portugal

e-mail: lopes@fe.up.pt

Static mixers are currently used in industrial applications to perform continuous operations and have become standard equipment since the 1970s for the mixing of miscible or immiscible fluids, homogenization of solid particles, and for the enhancement heat and mass transfer [1, 2], as an alternative to mechanically stirred vessels. The effectiveness of static mixers to deliver good mixing, or to enhance heat and mass transfer, derives from their capacity to statically induce radial mixing and to bring fluid elements into close proximity [2]. In these devices, products are mixed only by the energy of the flow, and they do not require external power, except for the necessary energy for pumping the fluids through the mixer, And this combined with small equipment size, low equipment cost and no moving parts [2].

The NETmix reactor is a new static mixer/reactor consisting of a network of mixing chambers interconnected by transport channels [3]. Networks are generated by the repetition of units of mixing cells, where each mixing unit cell consists of one mixing chamber, two inlets and two outlet channels oriented at a 45° angle from the main flow direction. Above a critical channel Reynolds number, the device main flow pattern evolves to a self-sustained oscillatory laminar flow regime inside the mixing chambers inducing local strong laminar mixing, which results from the geometrical features of the NETmix reactor. A network model was previously developed to describe and predict the behavior and mixing performance of NETmix [4]. From the point of view of modeling, and in practical terms, it was shown that chambers behave as perfectly mixing zones and the channels as plug flow segregation zones. The NETmix reactor can therefore be conceptualized in mixing terms, as essentially a plug flow reactor with local maximum mixing. The mixing degree was defined and quantified [4, 5] showing that mixing can be controlled effectively and efficiently making it particularly suited for complex and fast kinetics reactions. Furthermore, NETmix is also a good efficient device, for heat and mass transfer, due to its high surface to volume ratio of the order of 10^3 m^2/m^3, one or two orders of magnitude higher when compared to conventional heat exchanger/reactors [6], and conventional reactors for catalytic reactions [7, 8]. Although the surface to volume ratio is one order of magnitude smaller than for micro-structured reactors [8, 9], the heat and mass transfer coefficients are expected to be strongly enhanced by the induced laminar mixing oscillations in the NETmix chambers, above the critical Reynolds number, [10, 11], and additionally preforming lower pressure drops when compared with micro-reactors.

Computational fluid dynamics (CFD) is an increasingly effective alternative to speed up equipment design and to be used as a fundamental tool for optimizing heat and mass transfer using static mixers [1]. In this work, a new 3D CFD model is introduced that enables the computation of transport properties of a more realistic representation of the NETmix geometry, the *NETmix Unit Block*(NUB) model. This simulation tool can be applied to any particular geometry of the NETmix reactor and overcomes the need of experimental data each time a new configuration is designed. The NUB model was used to study heat and mass transfer in the NETmix reactor and from this work, correlations of unit cell Nusselt and Sherwood numbers were developed. The present study suggests that 3-5 times higher heat/mass transfer rates are achieved when compared to the efficient parallel plates devices. Maximum convective heat and mass transfer coefficients are achieved when the flow inside the mixing chambers evolves to a self-sustained oscillatory laminar flow regime inducing strong local laminar mixing. Finally, a phenomenological model is presented to assess the rates of interfacial area formation along the NETmix reactor and compared with VOF dynamic mixture CFD simulations with the NUB model.

[1] E.S. Szalai, F.J. Muzzio, Fundamental approach to the design and optimization of static mixers, AIChE Journal, 49 (2003) 2687-2699.

[2] R.K. Thakur, C. Vial, K.D.P. Nigam, E.B. Nauman, G. Djelveh, Static Mixers in the Process Industries—A Review, Chemical Engineering Research and Design, 81 (2003) 787-826.

[3] J.C.B. Lopes, P.E. Laranjeira, M.M. Dias, A.A. Martins, Network mixer and related mixing process. PCT/IB2005/000647, February 2005. European Patent EP172643 B1, October 2008, in, 2005.

[4] P.E. Laranjeira, A.A. Martins, J.C.B. Lopes, M.M. Dias, NETmix®, a new type of static mixer: Modeling, simulation, macromixing, and micromixing characterization, AIChE Journal, 55 (2009) 2226-2243.

[5] P.J. Gomes, C.P. Fonte, R.J. Santos, M.M. Dias, J.C.B. Lopes, Experimental and numerical characterization and quantification of mixing in a NETmix® reactor, in: ECCE 7 - 7th European Congress of Chemical Engineering & CHISA 2010- 19th internation Congress and Process Engineering, Process Engineerig Publisher, Prahe, Prague, Czech Republic, 2010.

[6] Z. Anxionnaz, M. Cabassud, C. Gourdon, P. Tochon, Heat exchanger/reactors (HEX reactors): Concepts, technologies: State-of-the-art, Chemical Engineering and Processing: Process Intensification, 47 (2008) 2029-2050.

[7] F.J. Keil, Catalytic reactions and reactors, Chemical Engineering Science, 59 (2004) 5473-5478.

[8] L. Kiwi-Minsker, A. Renken, Microstructured reactors for catalytic reactions, Catalysis Today, 110

(2005) 2-14.

[9] M.N. Kashid, L. Kiwi-Minsker, Microstructured Reactors for Multiphase Reactions: State of the Art, Industrial & Engineering Chemistry Research, 48 (2009) 6465-6485.

[10] J.G. Khinast, A. Bauer, D. Bolz, A. Panarello, Mass-transfer enhancement by static mixers in a wall-coated catalytic reactor, Chemical Engineering Science, 58 (2003) 1063-1070.

[11] D. Li, K. Xiong, Z. Yang, C. Liu, X. Feng, X. Lu, Process intensification of heterogeneous photocatalysis with static mixer: Enhanced mass transfer of reactive species, Catalysis Today, 175 (2011) 322-327.

(70e) The Simplest Stirred Tank for Laminar Mixing: Mixing in a Vessel Agitated By An Off-Centered and Angled Disc

Alvarez, M. M., Tecnologico de Monterrey
Bulnes-Abundis, D., Instituto Tecnologico y de Estudios Superiores de Monterrey
Originally presented on: 11/4/2013 9:50:00 - 10:10:00

We study the mixing structure, performance, and short term dynamics in round bottomed laminar tanks agitated by an eccentrically located angled disc. We define eccentricity (E=e/R) as the ratio of the distance of the axis of rotation from the center line of the tank (e) and the tank radius (R). The structural and dynamic features observed at different eccentricity values were compared using planar laser induced fluorescence (p-LIF) techniques and computational fluid dynamics (CFD) calculations. A PoincarÃ© analysis demonstrates the chaotic nature of the flow induced by eccentricity. Practically globally chaotic conditions are observed for E=0.42 and E=0.50, with mixing times of 5-8 minutes at Re=416. We study the effect of different injection points on the short term mixing dynamics and we calculate axial flow rates and Power numbers. Stirred tanks agitated by an eccentrically located angled disc are a simple and cost effective system for laminar mixing applications.

(70f) Low Power, Baffle-Free Mixing With Contra-Rotating Impellers

Register, J., U. South Carolina
Regalbuto, J. A.,
Regalbuto, J. R., University of South Carolina
Originally presented on: 11/4/2013 10:10:00 - 10:30:00

A novel method to achieve good mixing without baffles in tank reactors and other mixing vessels will be presented. Baffle free mixing can be achieved with contra-rotating impellers on coaxial shafts, driven by a gearbox connected to a single motor. Contra-rotating impellers can be run in a variety of configurations, such as the top impeller pushing down, bottom pushing up, or top pushing up and bottom down.

Mixing experiments were conducted in a 5 gallon cylindrical tank using a Caframo BDC 3030 Stirrer and standard pitched blade turbines. Torque was measured as a function of impeller speed and the degree of mixing was imaged with suspended colored particles. A single impeller, with and without baffles, was compared to dual impellers on the same shaft, with and without baffles, and then to baffle-free contra-rotating impellers in a number of configurations.

Good mixing was never achieved without baffles using single or dual impellers on a single shaft. The measured torque was proportional to the turbine blade area; dual impellers required about twice as much torque as a single impellers. Baffles approximately double the required torque for both of these cases at the conditions tested. Contra-rotating impellers have higher torque than single impellers because of the added blade area, but incur no torque penalty of baffles. Good mixing was achieved in the optimal contra-rotating configuration mixing using about 20% less torque, and at lower rotational speeds than the single or dual impellers with baffles on a single shaft. Thus, the power required for good mixing with contra-rotating impellers with no baffles is significantly less than single shaft mixing with baffles. A number of applications of this mixing apparatus will be discussed.

(70g) New Impeller Type for Significant Mixing Efficiency Improvement and Energy Saving

Wang, L.,

Originally presented on: 11/4/2013 10:30:00 - 10:50:00

A new type of impeller has been developed demonstrating significant mixing efficiency improvement and energy saving. The mixing mechanism and flow pattern produced are different from that of any other products in the market. The newly developed impeller vanes don't have surface perpendicular to the impeller rotating direction which produces vertex. Vertex flow provides little mixing and requires vessel baffles to break the vertex to improve the mixing and much more energy input is required. The new impeller produces true radial flow pattern. Liquid is discharged outward at various directions and flow back to the impeller from both the vessel bottom and liquid surface. The impeller forms a centre of the flow of liquid in a vessel. The velocity gradient is gently decreased from the impeller to the vessel body and liquid surface. The flow pattern is easy to model and calculate for industry applications. For solids suspension applications, it presents superior satisfactory results. Solids settled at the vessel bottom are shown to be self-cleaned. The impeller produces minimum vertex and no baffles to the vessel are needed so maximizes the energy output for mixing.

(132a) An Unusual State During the Production of Solid Stabilized Emulsions in Stirred Tanks: Paste

Ayranci, I., University of Alberta
Fradette, L., Ecole Polytechnique de Montreal
Originally presented on: 11/4/2013 12:30:00 - 12:48:00

An unusual state during the production of solid stabilized emulsions in stirred tanks: paste

Solid stabilized emulsions (SSEs) are alternatives to conventional surfactant based emulsions. In SSEs the particles get adsorbed at the oil-water interface irreversibly, giving a more stable emulsion than the conventional emulsions. This is advantageous when the production of an emulsion is targeted. It may, however, be disadvantageous if undesired emulsions form in the presence of the three phases ? oil-water-solids. An example of that is in oil sands processing. The small particles adsorbed at the oil-water interface form very small and very stable water droplets that are difficult to separate from the oil. In either case, whether the SSEs are the desired or the undesired product, there is a lack of knowledge of the necessary process conditions for the production. The current research on SSEs is limited to geometries that are not directly applicable to industry such as hand-shaken vials or cells stirred with homogenizers at overwhelmingly high Reynolds numbers, and to characterizing the emulsions. Our global objective is to demonstrate the impact of the process conditions on the production of SSEs in stirred tanks. This work focuses on an unusual state that the SSEs go through during production: paste state.

Under continuous mixing, when the three phases, oil and water-solids slurry, are brought together first a normal SSE forms. Shortly after, the emulsion turns into a paste of particles in oil and water. Depending on the process conditions the time it takes to reach the paste state can be very short, or very long. The paste is not a proper emulsion, but it is not a complete separation of three phases. This paste state is similar to the rag layer that is seen during oil sands processing. The rag layer consists of some small droplets, oil and water phases and sand. Over time, as mixing of the paste is continued the paste turns back into a normal solid stabilized emulsion. The specific objective of this work is to determine the parameters that affect the formation of the paste, and to analyze how they affect the formation, and the time it takes to reach a paste. Â

The salt concentration, particle diameter, wettability of the particles, solids concentration, hydrodynamics and Reynolds number were found to affect the formation of the paste and time required for pasting. The tests were run for two impellers, a Rushton turbine and a pitched blade turbine, at varying Reynolds numbers, with or without baffles (with an off-centred impeller).Â Phase diagrams that show the regions where paste forms for each impeller at varying oil, water and solids concentrations were generated. The emulsion formed after continued mixing of the paste also exhibited surprising behaviour by completely separating into three phases. The time required for this separation to occur varies based on the process conditions.

(132b) Macro-Mixing of the Dispersed Phase in An Immiscible Liquid-Liquid Stirred Reactor
Yang, C.,
Originally presented on: 11/4/2013 12:48:00 - 13:06:00

Macro-mixing of the dispersed phase in an immiscible liquid-liquid stirred reactor

Dang Cheng, Xin Feng, Jingcai Cheng, Chao Yang *

Institute of Process Engineering, Chinese Academy of Sciences,

Beijing 100190, China

Abstract

Liquid-liquid dispersions in mechanically agitated vessels are frequently encountered in chemical industry. The dispersed drops are continuously broken and coalesced, which make such processes extremely difficult to be well controlled. Numerous investigations have thus been devoted to the two-phase dispersed systems because of their practical importance and complexity [1]. Although they were widely studied in scientific and engineering circles, there is still little information regarding the mixing in either or both phases in stirred liquid-liquid dispersions. The knowledge of macro-mixing, which is usually globally characterized by mixing time (also termed as blend time), is essential when the mixing rate is comparable with or slower than the chemical reaction rate. Under these circumstances, the liquid-liquid mixing efficiency is found to have a profound impact on the overall performance, e.g., production costs and product selectivities.

Mixing is a very complex process in liquid-liquid dispersions, which can occur either in the continuous phase or in the dispersed phase or in both phases. Recently, Zhao et al. [2] investigated the macro-mixing of the continuous phase in an immiscible liquid-liquid stirred reactor using the conductivity technique, in which the tap water was taken as the continuous phase and several oil phases were used as the dispersed phases, and their study revealed the non-linear effect of the dispersed phase holdup on the continuous phase mixing time. The macro-mixing process in the continuous phase of a multiphase system is achieved by convective (bulk) and turbulent (eddy) motions. However, the problem of mixing amongst the dispersed phase (drops), i.e., the dispersed phase mixing, is of special interest for such practical applications as suspension and emulsion polymerization, i.e., reactions occurring in the dispersed phase, in which the dispersed phase mixing may greatly influence the productivity and selectivity. Experimental measurements and theoretical studies have shown that the dispersed phase mixing had a significant influence on the average reaction rates and product selectivities in non-first order or mass transfer controlled reactions occurred in the dispersed phase [3,4].

The dispersed phase mixing is fulfilled by means of intense drop interactions (dispersion, breakup and coalescence), which control the species concentration distribution among drop population. The drops are naturally completely segregated in liquid-liquid dispersions, if there is no dispersed phase mixing or there is no segregation (no concentration difference among drops) at an infinite dispersed phase mixing rate (or interaction rate) between the dispersed drops. Rietema [5] stated that the two extreme cases did not really exist in practical industrial processes and usually there was a finite interaction rate or partial segregation. Thus in many practical reactors the dispersed phase cannot be considered to be either totally unmixed or perfectly mixed. On analyzing the problem, there arises the question to what an extent the drop concentrations can be mixed and how long it usually takes for an added nontransferring tracer to be mixed to an expected extent. However, it is particularly noted that quantitative measurement of the macro-mixing process in the dispersed phase is still an unresolved problem, despite of its practical importance.

In our previous works [2,6,7], the macro-mixing and two-phase flow in the continuous phase in immiscible

(132c) The Effect of Particle Shape On the Scale Up of Just Suspended Impeller Speed

DiMattesa, M., Rowan University
Cendo, D., Rowan University
Boesch, S., Rowan University
Etchells, A. W. III, AWE3 Enterprises
Hesketh, R. P., Rowan University
Originally presented on: 11/4/2013 13:06:00 - 13:24:00

The goal of this study is to determine the minimum stirrer speed required to generate suspensions of spheres, rods, and plates. Most of the previous work that has been done on just suspended impeller speed has been done for spherical particles and there are only a small number of research studies on the effect of particle shape on the just suspended impeller speed. As a result of this uncertainty in the effect of particle shape on just suspended impeller speed it is estimated that industries use more energy than is required to reach this state. In this study experiments were conducted with 3 shapes of particles over a range in particle sizes and solid mass fractions. It was found that at equivalent solid-liquid mass fractions and particle volumes, spheres were the most difficult to suspend requiring about 1.1 times higher stirrer speeds than both rods and plates. In addition, the required speed to just suspend rods and plates of equal particle volume was very similar. Rods with diameters from 0.8 to 2 mm and lengths from 10 to 30 mms were tested, and it was concluded that the minimum dimension of the rod has the most significant impact on predicting just suspended impeller speed. Plates with various dimensions were tested, and it was concluded that both the side length and thickness have an impact on the just suspended impeller speed. Correlations of just suspended impeller speed and a volume/surface area shape factor show a maximum for plates and sphere at the 0.33 mm value, but further tests are required to confirm this behavior for rods. Scaled up testing was done using a geometrically similar tank that was 2.5 times larger than the small scale tank. It was concluded from these results that the best method of predicting scale up of just suspended speed for our system was to use a constant Froude number. This method resulted in an overall error of less than 3%.

(132d) Effect of Impeller Submergence, Position, Diameter and Speed On Flow Regime, Surface Air Entrainment and Mixing Effectiveness of Partially Filled Stirred Vessels

Motamedvaziri, S., New Jersey Institute of Technology
Armenante, P. M., New Jersey Institute of Technology
Originally presented on: 11/4/2013 13:24:00 - 13:42:00

In many industrial applications, mixing vessels have a liquid height-to-tank diameter ratio, H/T, equal to, or larger than 1. However, there are many instances where this ratio is lower than 1, as in all those cases in which the vessel is either emptied or filled. Even when H/T<1, sufficient agitation must still be provided in order to attain the desired process objectives. When the impeller submergence is reduced as a result of lowering the liquid level, the fluid dynamics of even a single-phase stirred liquid can become quite complex, with different regimes possibly existing depending on the geometric characteristics of the system (such as impeller clearance, liquid height, or liquid submergence above the impeller). The objectives of this work were to study in detail the hydrodynamic changes that occur when H/T is decreased, and to determine the minimum liquid levels and the critical impeller submergence for different impeller off-bottom distances, impeller diameters and agitation speeds where adequate mixing process can still be achieved, both in a single liquid phase and in solid-liquid suspensions. A number of experimental tools were used to analyze the systems under investigation, including: Particle Image Velocimetry (PIV) for the experimental determination of the velocity profiles, and after the appropriate data manipulation, the impeller pumping capacity and its Pumping Number; a strain gage-based rotary torque transducer system to measure the power dissipated by the impeller; and a visual observation method to determine the minimum agitation speed for complete solids suspension, N_{js}. In addition, Computational Fluid Dynamics (CFD) modeling was used to predict the behavior of the system in terms of its velocity profile, Power Number, and Pumping Number. CFD predictions were obtained using a multiple reference frame (MRF) model coupled, when needed, with a Volume of Fluid (VOF) model, in order to study systems in which a vortex could be expected to form.

In general, good agreements between the experimental data and the predicted results for the velocities distribution, Power Number, radial Pumping Number, mixing time, and N_{js} were obtained. Results show that there is a critical impeller submergence ratio S_b/D below which: (1) the macroscopic flow pattern generated by the impeller changes substantially, transitioning from either a "double-loop" recirculation flow or a "single-loop" recirculation flow (depending on the impeller clearance off the tank bottom) to an upward "single-loop" recirculation flow; (2) the Power Number and radial Pumping Number drop significantly; (3) solid suspension cannot be attained at any agitation speed; (5) vortex formation occurs, air entrainment is significantly facilitated, and impeller flooding typically results. This is the first time that such hydrodynamic regime change has been reported and characterized. When this flow regime transition occurred, it was observed that the average velocity field and turbulence intensity close to the tank bottom decreased substantially; this was identified as the reason why solid suspension became unachievable.

Furthermore, the critical impeller submergence ratio resulting in the establishment of the newly described flow regime was not affected by dynamic variables such as agitation speed. Impeller flooding was observed only when the new regime was established and when the vortex depth reached the impeller disk. This phenomenon was correlated to critical values of the Froude number. Additionally, by decreasing the impeller off-bottom clearances the operating window within which the stirred vessels could be operated effectively became wider. However, for low impeller off-bottom clearances ($C_b/T<0.05$) and low impeller submergence ratios ($S_b/D<0.37$, corresponding to $H/T<0.16$) even the lower recirculation loop was suppressed (tank bottom effect) and an unstable flow system was observed with no clear recirculation, implying that effective mixing could not achieved.

(132e) Drop Breakage By Shear Tip Speed

Rodgers, T. L., The University of Manchester
Cooke, M., University of Manchester
Originally presented on: 11/4/2013 13:42:00 - 14:00:00

Many products produced in the process industries involve emulsification where the drop size distribution affects the processing and the product properties, e.g. food, cosmetic, pharmaceutical, and health care products. For example, in a two-phase process, the mass transfer rate between the phases is proportional to the interfacial area. This interfacial area changes with the drop size distribution, which varies with the conditions inside the vessel and time. Therefore, an understanding of the mechanisms for drop break-up is key for any design process. For inviscid systems, there are two main mechanisms that are thought to break drops: break-up due to turbulent eddies, i.e. energy dissipation rate; and break-up due to the agitator shear rate.

Break-up due to turbulent eddies is generally based on the work of Kolmogorov (1949) and Hinze (1955) which utilises the concept of eddy turbulence to define a limiting drop size. It is usually assumed that drop break-up occurs due to the interactions of drops with the turbulent eddies of sufficient energy to break the drop. Therefore, for a given fluid system the effective equilibrium drop size (this is the drop size after a sensible processing time, when the drop size reduction with time is very small and almost unmeasurable) is dependent on the energy per unit mass and thus should scale-up with this value when using geometrically similar vessels.

Break-up due to the agitator shear rate is based on a balance between the external viscous stresses and the surface tension forces (Liao and Lucas, 2009). If the break-up is due to the agitator shear rate then the effective equilibrium drop size is related to the maximum shear rate. This would mean that lower power number agitators can produce smaller drops than higher power number agitators, as low power number agitators may have a higher shear rate, which has been seen experimentally by Zhou and Kresta (1998).

This paper presents the shear tip speed, KsND, and shows it how drop size data for a wide variety of agitators can be correlated against the shear tip speed, which is proportional to the maximum shear rate generated by the agitator. This paper continues with analysis of the drop sizes produced by a variety of turbines and modified turbine impellers showing better correlation with the shear tip speed.

Hinze, J. O. 1955. AIChE Journal 1, 289-295.

Kolmogorov, A.M. 1949. Doklady Akademii Nauk 66, 825-828.

Liao, Y. and Lucas, D. 2009. Chem. Eng. Sci. 64, 3389-3406.

Rodgers, T. L. and Cooke, M. 2012. Chem. Eng. Res. Des., 90, 323-327.

Zhou, G. and Kresta, S. M. 1998. Chem. Eng. Sci. 53, 2063-2079.

(132f) A Process-Based Analysis On the Generation of Pickering Emulsions

Tsabet, E., NSERC/TOTAL Industrial chair, Montreal Polytechnic
Fradette, L., Ecole Polytechnique de Montreal
Originally presented on: 11/4/2013 14:00:00 - 14:18:00

The replacement of surfactants by solid particles in emulsification processes offers many advantages and raises new opportunities. Unfortunately, the current state of knowledge does not allow for process design mainly because the lack of information about the stabilization mechanisms which are completely different from surfactant-based systems. At a process level these mechanisms can be divided into four steps: Droplet formation by breakage, the particle/droplet approach and collision, the particle adsorption and the formation of the particles network. Being mainly interested in the development of new emulsification processes, the global aim of this work is to model the generation and the stabilization of the droplets with particles in a system representative of industrial conditions. This work specifically investigated the impact of process parameters on the resulting emulsions.

Emulsification experiments were performed in an unbaffled tank using an off-centred pitched blade turbine. The parameters investigated were time under mixing, particles fraction, impeller speed and oil viscosity. Emulsification efficiency was quantified by means of size distribution measurements (Malvern Mastersizer 3000). Visualization techniques (PVM probe from Mettler Toledo and Optical microscopy) were also used to observe emulsion evolution during processing and evaluate the final droplet coverage by the particles.

Results showed that all the parameters affecting the hydrodynamics interact strongly and optimal conditions should be defined to produce the smallest droplets with the narrowest distribution. This interaction can be attributed to the fact that two different phenomena are involved simultaneously: droplet production and droplet coverage. This is completely different from surfactant-based systems where the stabilization step is much faster than the droplet production in addition to promoting droplet breakage by reducing the interfacial tension. Impeller speed and the emulsification time results highlighted the relevance of the fluid circulation and process time in the stabilization mechanism. The particles fraction results clearly showed the importance of the stabilization rate.

(132g) The Taylor Dispersivity of a Passive Tracer in the Pressure-Driven Flow of a Concentrated Suspension of Rigid, Non-Colloidal Spheres

Ramachandran, A., University of Toronto

Originally presented on: 11/4/2013 14:18:00 - 14:36:00

It is well known that the axial spreading of a slug of passive tracer in a Newtonian fluid flowing in a conduit under laminar conditions is governed by a Taylor dispersivity[1], which scales as U^2B^2/D. Here U is the average velocity of the fluid in the conduit, B is the characteristic depth of the conduit, and D is the molecular diffusivity. The Taylor dispersivity is directly proportional to characteristic equilibration time, t_c, of the solute concentration inhomogeneities over the cross-section. The scaling of the Taylor dispersivity may be viewed as $U\,(Ut_c)$, where t_c is the diffusion time scale over the characteristic depth of the channel, $B^2/\,D$. The shorter this time scale, the larger the Taylor dispersivity, and the faster the axial spreading of the solute.

In this work, the Taylor dispersivity of a solute being carried in the suspending fluid of a concentrated suspension of rigid, non-colloidal particles is analyzed. The model for solute transport employs the phase-averaging approach adopted by Zydney and Colton[2], but incorporates a tensorial self-diffusion tensor based on experimental measurements. Particle migration effects are included via the suspension balance model of Nott and Brady[3] along with the constitutive equations of Zarraga and Leighton[4].

For suspensions, in addition to molecular diffusion, there are two more mechanisms influencing the characteristic equilibration time, t_c: shear-induced self-diffusion[5], and secondary currents driven by second normal stress differences. As the effects of shear-induced self diffusion and secondary currents become stronger, t_c, and therefore the Taylor dispersivity are decreased. Regimes where each of the three mechanisms can provide a dominant contribution to the Taylor dispersivity are identified in terms of three dimensionless parameters: $B^2/a^2g(\varphi)$, $k\delta(\varphi)$ and Pe$=UB/D$. Here a is the particle radius, $g(\varphi)$ represents the dependence of shear-induced migration rate on particle volume fraction φ, $\delta(\varphi)$ is the reduced second normal stress difference N_2/τ, Pe is the Peclet number, and k is a constant related to the shape of the cross-section. The scaling for Taylor dispersivity is established in each regime, and is verified with numerical computations.

The theoretical results developed in this work will allow us to interpret experimental measurements of Taylor dispersivity to deduce the magnitudes of shear-induced solute self- diffusivities at various volume fractions, and to measure the contribution of secondary currents in solute transport. This work is a step towards a more general understanding of how solute is transported in more complicated suspensions, such as blood.

References

1. G. Taylor, Dispersion of soluble matter in solvent flowing slowly through a tube, *Proc. R. Soc. Lond. A* **219**, 186-203 (1953).
2. A. L. Zydney and C. K. Colton, Augmented solute transport in the shear flow of a concentrated suspension, *Physicochem. Hyd.* **10**, 77?96 (1988).
3. P. R. Nott and J. F. Brady, Pressure-driven flow of suspensions: simulation and theory, *J. Fluid Mech* **275**, 157-199 (1994).
4. I. E. Zarraga, D. A. Hill and D. T. Leighton, The characterization of the total stress of concentrated suspensions of noncolloidal spheres in Newtonian fluids., *J. Rheol* **181**, 415-439 (1987).
5. E. C. Eckstein, D. G. Bailey and A. H. Shapiro, Self diffusion of particles in shear flow of a suspension, *J. Fluid Mech.* **79** 191?208 (1976).

(132h) Improving Euler-Granular Models in Multiphase Mixing for Polydispersed Flows

Rosendall, B., Bechtel National, Inc.

Originally presented on: 11/4/2013 14:36:00 - 14:54:00

The purpose of this presentation is to evaluate computational fluid dynamics modeling for granular flows for polydisperse solids and their motion within multiple carrier fluid systems. Much of the established work is based on monodisperse particles within a single-carrier fluid system, and their extension requires special attention. Here, a consistency constraint is proposed as a basic test of such an extension. Based on this constraint, it is found that some of the most common models fail. These models are adapted to pass the consistency criteria, and applied to a number of test cases that indicates that relatively small model reformulations can enable models established in a monodisperse solids flow in a single carrier fluid, to produce accurate results in highly polydisperse particle flows with multiple carrier fluids. The model reformulations are applied to a polydisperse granular flow in a bladed mixer (Remy, *et. al.* 2011).

(188a) Perspectives On the Use of Simulation and Correlation for Design of Mixed Chemical Reactors

Patterson, G. K., Missouri University of Science and Technology

Originally presented on: 11/4/2013 15:15:00 - 16:15:00

The last three decades have seen the development of sophisticated methods for simulating flow and mixing for application to the design and scale-up of mixed chemical reactors. Alas, the best, most accurate methods for simulation consume much computer time, even with today's large-memory, fast computers. This paper explores ways to develop approximate methods to efficiently simulate flow, mixing and chemical reaction and use data correlation to support design and scale-up. Work over these last three decades will be reviewed to show how these less demanding methods have been developed. Future work will be suggested for further refining the methods discussed and determining where their inaccuracies occur and how they may be addressed.

(248a) Synthesis of Fine Chemicals by Heterogeneous Catalysis: A Critical Approach of the Role of Solvent for Hydrogenation Reactions

Apesteguia, C.,

Originally presented on: 11/5/2013 8:30:00 - 8:55:00

The synthesis of fine chemicals over solid catalysts often involves the use of solvents that may strongly influence the catalyst performance. Thus, the choice of suitable solvents is frequently critical to obtain high catalytic activity and selectivity. However, the optimal solvent selection requires a detailed knowledge on the relationship between the chemical nature of the solvents and the interactions taking place in the gas-liquid-solid catalytic systems. One of the most common types of catalytic reactions carried out in the presence of solvents is the hydrogenation of organic compounds. In particular, the selective hydrogenation of aromatic ketones into the corresponding alcohols on metal-based catalysts in the presence of different solvents has been widely studied. Here, the solvent effect on catalyst activity and selectivity for the liquid-phase hydrogenation of acetophenone to 1-phenylethanol was thoroughly investigated over Ni/SiO_2. Solvents of different properties and polarities were used: i) protic solvents: methanol, ethanol, 1-propanol and 2-propanol; ii) aprotic polar solvents: acetonitrile, g-butyrolactone and tetrahydrofuran; iii) aprotic apolar solvents: cyclohexane, toluene and benzene. The relative interactions solvent–catalyst, solvent–reactant and reactant–catalyst and their influence on the activity pattern were considered in the analysis. Concerning the influence of the solvent-reactant interaction, classical polarity parameters (e.g. dipole moment m and dielectric constant ε) and other solvatochromic scales (e.g. hydrogen-bond donor (a) and hydrogen-bond acceptor (b) parameters, p* polarity/polarizability index, Kosower's Z and ET(30) scales) were taken into account. The influence of H_2 solubility in the solvents was also considered. For the solvent-catalyst and reactant-catalyst interactions, the corresponding molar adsorption enthalpies were measured calorimetrically and compared.

(259a) Laminar Flow Pattern and Heat Transfer in a Vessel With Eccentrically Positioned Multiple Inclined Blade Impellers

Koot, M., Delft University of Technology
Van den Akker, H. E. A., Delft University of Technology
Originally presented on: 11/5/2013 8:30:00 - 8:50:00

The laminar flow pattern and the heat transfer from vessel wall to liquid in a vessel provided with three inclined blade impellers mounted on a single vertical shaft have been studied computationally. The axis with the three impellers is at an eccentric position and the flow is certainly laminar due to the high viscosity of the liquid.

The Computational Fluid Dynamics (CFD) simulations have been carried out with the help of the commercial flow solver ANSYS 14.5. Over 420,000 grid cells were used, almost 170,000 of which were in the impeller zone. To accommodate the revolving impeller, two options have been explored: the Multiple Reference Frame (MRF) option as well as the Sliding Mesh (SM) option, the former resulting in a steady-state solver while the latter is transient.

The flow patterns found with these two options are very different: only the SM option predicts a single circulation over the full vessel height in the part of the vessel opposite to the impeller position. Of course, the heat transfer from vessel wall to highly viscous liquid strongly depends on the flow pattern.

(259b) The Effects of a Spatially Variant Velocity Field On Stretching: Intuitive Measures

Bigio, D.,
Nixon, J.,
Originally presented on: 11/5/2013 8:50:00 - 9:10:00

Laminar mixing theory describes the 'goodness' of mixing as a function of increased surface area shared between two fluid species. Previous work in the field attempts to describe mixing from a post processing perspective, namely as a function of stretching history, while disregarding the nature of the underlying flow. In this work, new mathematical measures are derived which allow for the prediction of fluid body interface behavior in a flow and the exploration of the underlying flow fields itself. This family of measures creates a more intuitive basis for the exploration interfacial growth, providing a new perspective on laminar mixing. One set of measure relates velocity with the principal directions while the second set relates interfacial orientation with the princiapl directions. To explore the usefulness of these new measures, they are simulated in three numerical flows, steady shear, steady divergent flow, and spatially variant lid driven cavity.

In these flow geometries, the new family of measures proves valuable for demonstration of various features underlying flow including growth regime characteristics, transition in growth regime, as well as other more flow characteristics unique to each flow type. It has been shown that the change of mixing consistent with reorientation occurs after a rapid change in the relationship of the flow and principle directions. The second set of measures in this family allows for interface growth and modeled to be studied more objectively.

(259c) Measuring Blend Time In Laboratory Agitated Vessels: A Case for the Acid-Base-Indicator Method

Fasano, J. B., Mixer Engineering Company
Myers, K. J., University of Dayton
Janz, E. E., Chemineer, Inc.
Originally presented on: 11/5/2013 9:10:00 - 9:30:00

This paper will examine the sources of variance that can occur in the measurement of blend time. Source of variance in blend time measurements can result from human error, statistical variance, or instrument accuracy. These various sources of variance will be discussed as they relate specifically to the acid-base-indicator method. Discussions will include the ability of the human eye to see color change and at what pixel level. The video record/analysis technique will also be examined in terms of color and pixels.

The "partial blend time" technique will be described and how final blend time is determined. Paper references will be provided to substantiate this "partial blend time" procedure.

A visual perspective of the acid-base technique along with techniques to minimize human error will be provided. Assumptions and calculations used to determine the approximate degree of uniformity will also be submitted. Details and logic for utilizing the partial blend time technique will be presented.

Pros and cons of this technique will be discoursed.

(259d) The Confined Impeller Stirred Tank: Blend Time

Machado, M. B., University of Alberta
Kresta, S., University of Alberta
Originally presented on: 11/5/2013 9:30:00 - 9:50:00

The dispersion and dissolution of chemical additives into liquids has to be efficient in order to achieve reasonable industrial yields. Current bench scale tests do not represent properly industrial scale behavior, and turbulence inhomogeneity in bench scale stirred tanks can lead to significant uncertainties about the process mixing requirements. In order to better represent the industrial scale, a new bench scale stirred tank has been developed: the Confined Impeller Stirred Tank (CIST). The CIST provides more uniform intensity of mixing over the volume of the tank and uses a much smaller volume of sample than the traditional stirred tank. The geometry of the CIST allows a more uniform turbulence distribution because the tank is filled with several impellers and therefore the portion occupied by the bulk of the tank is dramatically decreased. The CIST is a narrow and tall tank (T=7.6 cm, H=3T). Three sets of impellers have been used: the A310, Rushton and Intermig. Sets of five (Rushton and A310) or six (Intermig) impellers of a relatively large diameter (D=T/2 for Rushton and A310 and D=2T/3 for Intermig) were used. The current work assessed the blend time in the CIST using a spectrophotometer. The Reynolds number was varied from 2 000 to 40 000.

(259e) Mixing Time in Partially Filled Stirred Vessels

Motamedvaziri, S., New Jersey Institute of Technology
Armenante, P. M., New Jersey Institute of Technology
Originally presented on: 11/5/2013 9:50:00 - 10:10:00

Mixing time, the time required to achieve a predefined level of homogeneity of a tracer in a mechanically stirred vessel, is an important parameter to evaluate the mixing efficiency of the system. In this work, mixing time was obtained for the case in which the liquid height-to-tank diameter ratio, H/T, is less than 1, i.e., for partially filled vessels. The time, t_{95}, required to achieve a 95% level of tracer homogeneity a stirred vessel with a disk turbine with D/T=0.31, C_b/T=0.30 and for different H/T values (corresponding to different values of the impeller submergence ratio, S_b/D) was measured experimentally using a colorimetric technique coupled with image processing and predicted computationally using CFD simulations. In addition, a correlation between mixing time and t_{BR}, defined as the ratio between the liquid volume in the tank and the flow rate generated by the impeller, was developed. The experimental results show that there is a critical impeller submergence ratio S_{bc}/D below which mixing time increases suddenly and significantly (implying that the mixing efficiency drops significantly) compared the mixing time at higher submergence ratios. Above the critical impeller submergence ratio t_{95} decreases with decreasing impeller submergence ratio although the non-dimensional mixing time, $t_{95}N$, remains constant irrespective of the agitation speeds for each impeller submergence ratio. Lower values for $t_{95}N$ can be observed for lower submergence ratios. In general, good agreement between the experimental data at different S_b/D ratios, the predicted results from CFD simulations, and the proposed correlation for mixing time was obtained.

(259f) Impeller Placement Optimization; Mixing Versus Mechanical Shaft Fatigue

Lee, S. J., Philadelphia Mixing Solutions, Ltd.
Higbee, R., Philadelphia Mixing Solutions Ltd.
Wu, B., Philadelphia Mixing Solutions Ltd
Originally presented on: 11/5/2013 10:10:00 - 10:30:00

Appropriate mixing system design is a balance between performance and cost. For most mixing systems, all flow in the mixing vessel is induced by the impeller which creates predominantly symmetrical circulatory loops and, on average, does not produce a net horizontal flow impinging against an agitator assembly. However, in some applications a fluid inlet is placed adjacent to an impeller which subjects the impeller to a continuous flow oriented perpendicular to the impeller axis of rotation. Such side flow adversely affects the life of an agitator assembly due to fatigue loading. In a particular commercial waste water treatment mixing application, there was a desire to place an impeller in a high side flow inlet region of a basin which would have necessitated an unreasonably large shaft diameter to prevent premature shaft fatigue failure. Using a combination of CFD flow analysis and fatigue based shaft design; the impeller was placed at an appropriate height to both minimize the fatigue affects of the horizontal inlet flow, as well as to ensure proper mixing. Three separate CFD studies are presented – The originally requested configuration (impeller next to side flow), impeller situated as high in the vessel as possible (good fatigue life but poor mixing) and the final optimum configuration (acceptable fatigue life and acceptable mixing). Constant Bernoulli side flow forces were computed from time averaged constant flow velocities determined by the CFD studies which allowed the computation of mean and alternating force components whose frequency of application equaled the shaft rotations per minute. A Goodman fatigue analysis approach was utilized.

(259g) Simulations of Mobilization of Bingham Layers in Agitated Tanks

Derksen, J., University of Alberta

Originally presented on: 11/5/2013 10:30:00 - 10:50:00

Simulations of mobilization of Bingham layers in agitated tanks

Numerical simulations were used to study mobilization of a bottom layer of Bingham liquid by agitating a Newtonian liquid above the Bingham layer. The agitation is done by a pitched-blade impeller at a Reynolds number of 6000. The Bingham liquid and the Newtonian liquid are miscible. The parameter space of the simulations has a Bingham number Bn (defined as the ratio of yield stress over inertial stress) and a Richardson number Ri as dimensionless variables. The Richardson number quantifies the role of the density difference between the two liquids. The simulation procedure is based on the lattice-Boltzmann method for the flow dynamics, and a finite volume scheme to solve the local and time dependent composition of the liquid mixture. The moderate Reynolds number allows us to directly simulate the flow, without the use of turbulence closure or subgrid-scale models.

The simulations show how the bottom layer gets eroded by the turbulent flow above it (see the figure), and show that there exists a critical value of the Bingham number beyond which the turbulent flow is not able to remove the Bingham layer (compare the left and right panel of the figure). The Richardson number is less important for the mobilization process, but has impact on the time it takes for homogenization of the liquids after mobilization has occurred.

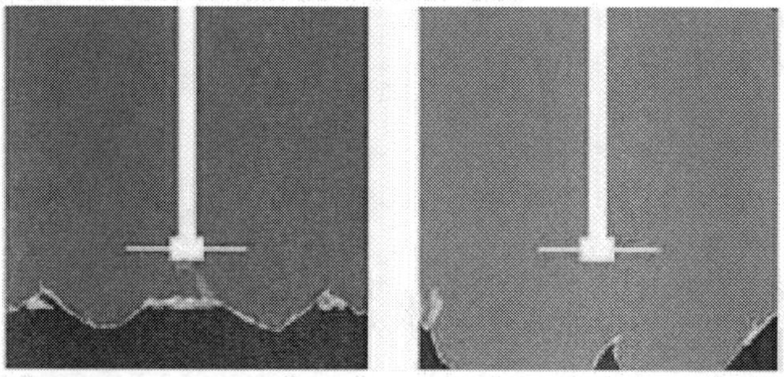

Figure: cross section through the mixing tank 100 impeller revolutions after start-up; left Bn=0.8, right Bn=0.1; blue indicates the Bingham liquid.Â

(316a) Formation of Poly(ε-caprolactone) Nanoparticles in Microstructured Mixers Via Solvent Replacement Precipitation: Fundamental Studies and Scale-Up Concepts

Kolbl, A., Karlsruhe Institute of Technology
Qayyum, A.,
Kraut, M., Forschungszentrum Karlsruhe GmbH
Dittmeyer, R., Karlsruhe Institute of Technology
Originally presented on: 11/5/2013 12:30:00 - 12:55:00

Poly(ε-caprolactone) nanoparticles are frequently employed in pharmaceutics for controlled drug release [1] and are often obtained via solvent replacement precipitation [2]. With microstructured mixers these particles were obtained free from mixing influences during their formation (size app. 100 - 500 nm). This is shown by two different approaches: 1.) the variation of the feed rates shows no influence on the particle size distribution. 2.) precipitation with different mixers (cyclone mixers [3] and multilamination mixers [4]), which could be clearly distinguished regarding their 'mixing qualities' using the iodide iodate reaction method, yield precipitates with the same particle size distribution. Hence, mixing in these microstructured mixers is significantly faster than the particle formation rate.

Fundamental studies:

Ruling out inhomogeneities caused by insufficient mixing, influencing factors on the particle size were examined using a microstructured cyclone mixer. With this mixer excellent mixing is obtained at low flow rates. This suggests that microstructured cyclone mixers are an effective tool for scientific research, especially when materials are expensive or difficult to obtain. Influencing factors which were identified in previous studies [2] are substantiated, namely the polymer outlet concentration and the composition of the solvent (acetone and water in different ratios). Examinations free from mixing influence allow for the quantification of these factors. It could be shown that the reproducibility increased inversely to particle size.

Scale-up concepts:

Mixing in multilamination mixers (V type and Y type) is significantly slower compared to mixing in cyclone type mixers but still significantly faster than the particle formation rate. IMVT's multilamination mixers are constructed from stacks of micromachined foils. By stacking micromachined foils a multichannel system is obtained. Scaling of those mixers is possible by the variation of the number of micromachined foils. Constant 'mixing qualities are obtained keeping a constant linear fluid velocity in the microchannels. The scaling concept has been investigated using the iodide iodate reaction method [4].

[1] R. Langer, Science 249 (1990) 1527-1533.

[2] F. Lince et al., Journal of Colloid and Interface Science 322 (2008) 505–515.

[3] A. Kölbl et al., Chemical Engineering Journal 167 (2011) 444–454.

[4] A. Kölbl et al., Chemical Engineering Journal 160 (2010) 865–872

(316b) Bioreactor Simulation With Cuda

Witz, C.,
Tantikul, T., Graz University of Technology
Khinast, J. G., Graz University of Technology
Originally presented on: 11/5/2013 12:55:00 - 13:20:00

Bioreactor simulation with CUDA

Christian Witz[1], Tawan Tantikul[1], Johannes Khinast[1,2]

[1]Graz University of Technology, Institute for Process and Particle Engineering

[2]Research Center Pharmaceutical Engineering

Abstract

The effectiveness of a biopharmaceutical manufacturing process depends to a large extent on the efficiency of the bioreactor, especially in the field of generic drugs. So far, the engineering process of the reactor design has been mostly driven by empirical knowledge, as the simulation of this complex multiphase and multiscale process was impossible for many years. However, despite the recent improvement of computational capabilities, the simulation of an industrial scale reactor takes months for only a few seconds of real operation time.

Thus, the goal of this study is to use graphic cards to speed up this simulation. The Compute Unified Device Architecture (CUDA) technology of nVidia has made the computational power of graphic processing units (GPUs) available for scientific calculations [1]. In the multiphase simulation, the large number of computing units in the GPU leads to a significant reduction of calculation time. To archive this acceleration, an efficiently parallelizable simulation method is needed.

The lattice Boltzmann method (LBM) which was developed based on the lattice gas automata [2], represents an efficient way to numerically capture the liquid phase flow dynamics on the GPU computing machine. It uses a regular grid with evenly distributed nodes. To model the geometry without changing the regular grid the immersed boundary method is used. The air bubbles are simulated with the Lagrangian particle tracking (LPT) method. The sum of the forces acting on each bubble, i.e. the drag, the buoyancy, the lift force, the history force, the added mass effect and gravity is used to determine the acceleration of the particle. The acceleration and the time step length give the velocity and the position change at the end of the time step.

For coalescence and breakup stochastic models are used. The phases are coupled with a two way approach.

As the calculation results are computed on the graphic card, simulation data can be visualized real time during the calculation. Furthermore to simulate large reactors, the code has a multi GPU functionality, hence, it can distribute the workload of the simulation on several graphic processors. This multi GPU code is able to handle the large scale, memory demanding simulation such as the simulation of the industrial scale reactor.

References

[1] http://www.nvidia.com/cuda

[2] Sukop, M.: Lattice Boltzmann Modeling, Springer, Berlin, 2007.

Submission details

Session - 26013 Mixing Issues and Scale-Up/Scale-Down Strategies for the Production of Pharmaceutical and Biopharmaceuticals

Emails:

christian.witz@tugraz.at
tawan.tantikul@tugraz.at
khinast@tugraz.at

(316c) Characterization of Kla Using Definitive Screening Designs for Fermentation Development and Scale-Up

Ornek, D.,
Anderson, R.,
Chopra, M.,
Lin, W.,
Ramsey, P.,
Originally presented on: 11/5/2013 13:20:00 - 13:45:00

Characterization of k_La using Definitive Screening Designs for Fermentation Development and Scale-Up

Fermentation development and optimization are typically performed at small scale to reduce their cost and time requirement. High cell density aerobic fermentation is scaled-up often based on keeping volumetric mass transfer coefficient (k_La) constant. To facilitate a successful transfer of an optimized small-scale fermentation process to lab, pilot, and industrial scale, the k_La profiles for these fermentors should be determined. Although a number of correlations and predictive models for k_La are available, they do not have universal applicability.

The purpose of this study was to use a statistical Design of Experiments (DoE) approach to characterize k_La in 1.3L DASGIP, 10L Biostat B, 40L Biostat CDCU and 200L ABEC microbial fermentors, and hence to use this information for fermentation scale-up. Polarographic dissolved oxygen sensors were used to measure dissolved oxygen concentrations in the fermentors. k_La was determined in de-ionized water by using the dynamic re-oxygenation method. Definitive Screening Designs, a new efficient statistical Design of Experiment, were used to characterize k_La in the fermentors. Four experimental factors, temperature, agitation, air flow rate, and volume, were included in the experimental design as input parameters, and k_La was the output parameter. Definitive Screening Designs showed that agitation, air rate and volume were statistically significant for all the fermentors. Air and volume had strong positive correlations with k_La, and agitation had strong quadratic effects on k_La. Results showed that four fermentors have comparable k_La profiles. The correlations determined here will be used as an effective tool for fermentation development, scale up and tech transfer.

(316d) Scaling Down the Production of Biopharmaceuticals to Continuous Micro-Chips

Alvarez, M. M., Tecnologico de Monterrey
Carrillo-Cocom, L. M., Tecnologico de Monterrey
Garza-Garcia, L. D., Tecnologico de Monterrey
Tapia-Mejia, E. J., Tecnologico de Monterrey
Camacho-Leon, S., Tecnologico de Monterrey
Rodriguez-Gonzalez, C. A., Tecnolgico de Monterrey
GarcÃa-Lopez, E., Tecnolgico de Monterrey

Originally presented on: 11/5/2013 13:45:00 - 14:10:00

We report the use of continuous flow micro-devices for the production of a glycosylated pharmaceutical product (a monoclonal antibody). We culture CHO cells on the surface of different PMMA/PDMS micro-channel chips. In particular, three different channel configurations were analyzed: (a) an ondular channel design, (b) a "zigzag" shaped micro-channel, and a (c) series of shallow compartments connected with a chanel. The surfaces of all micro-channels were textured with semi-spherical cavities and coated with fibronectin. Experiments of continuous perfusion at different flow rates were performed in the range of 0.1 to 6.0 microL/min. The velocity field at the different flow rates and geometries was investigated through CFD. Significant differences in attached cell density, cell proliferation, and monoclonal antibody production were observed among the different experimental scenarios. In some configurations, we observed sustained monoclonal antibody productions resulting in a steady-state productivity above 6,240 mg L^{-1} day^{-1}, at least 100 times higher than that observed in commercial scale fed-batch systems. We discuss the effect of surface texture, system geometry and flow rate on the performance of these micro-bio-reactors.

(316e) Computing Dissolved Oxygen Profiles in Aerobic Fermenters

Kasat, G., Tridiagonal Solutions Inc.
Vedapuri, D., Tridiagonal Solutions Inc.
Benz, G., Benz Technology International, Inc.
Originally presented on: 11/5/2013 14:10:00 - 14:35:00

Aerobic fermenters have been a staple method of bioprocessing for many years. Oxygen is necessary for all aerobic fermentation. Maintaining the appropriate and uniform concentration of dissolved oxygen (DO) is important for the successful growth of culture. Large DO gradients can lead to operating problems. Low or High DO levels in some part of reactors may lead to bi-product formation. A priori information on DO distribution can help in proper design and scale-up of industrial fermentation process.

Until now, there has been no way to reliably predict DO distribution. Indeed, it has not been possible to even measure it. A typical aerobic fermenter has only one DO probe; a few rare fermenters have two DO probes: one near the top of the fermenter, and one near the bottom. A substantial DO gradient has been observed in some fermenters: up to 100:1 ratio of the bottom DO to top DO with multiple radial turbines and up to 2:1 with axial/radial combinations.

Computational Fluid Dynamics (CFD) models have been successfully used to predict mixing, gas hold-up distribution, mass transfer & heat transfer rates in stirred gas-liquid fermenters. In this study the CFD models were used to predict DO distribution in a fermenter. Eulerian-Eulerian multiphase model along with standard k-e turbulence model has been used to predict the turbulent gas liquid flow in the vessel. The results of the CFD model were used to calculate oxygen transfer rates and DO distribution in the vessel. While calculating DO it has been assumed that the oxygen consumption rate (OUR) is uniform throughout the vessel.

Multiple simulations were carried to understand the impact of agitator design and operating condition on the DO distribution. Since the actual DO distribution cannot be experimentally verified, the absolute magnitudes of the results may not be accurate. However, it can be said that CFD models can be successfully used to predict DO distribution profiles in aerobic fermenters.

(316f) Hydrodynamic Effects of An Arch-Shaped Fiber Optic Probe in a USP Dissolution Testing Apparatus 2

Zhang, Y., New Jersey Institute of Technology
Bredael, G., Merck & Co., Inc.
Armenante, P. M., New Jersey Institute of Technology
Originally presented on: 11/5/2013 14:35:00 - 15:00:00

Dissolution testing is widely used in the pharmaceutical industry to evaluate newly developed drug formulations and as a quality control method to insure that solid dosage forms have consistent dissolution property. Typically, samples are manually drawn from the dissolution vessel prior to analysis. An approach to overcome the limitations of manual sampling consists in the use of sampling probes, such as fiber optic probes, permanently inserted in the dissolution medium and continually sampling the drug concentration in it as the solid dosage form dissolves. Despite their advantages, permanently inserted fiber optic probes can alter the normal fluid flow within the vessel and produce different dissolution testing results.

In this study, the hydrodynamic effects introduced by an arch-shaped fiber optic probe in a USP Dissolution Testing Apparatus 2 are studied by: (1) conducting dissolution tests, with and without the probe, using Prednisone tablets fixed at nine different locations at the bottom of the vessel and comparing the dissolution profiles obtained using statistical tools; and (2) experimentally determining the velocity profiles in the vessel, with and without the probe, using Particle Image Velocimetry (PIV) and quantifying changes in the flow velocities on selected horizontal iso-surfaces.

The results show that the arch shaped fiber optic probe does have a baffling effect on the hydrodynamics in the dissolution vessel. This effect results in changes in the velocities in the fluid flow, and therefore in changes in the dissolution rate of the tablets undergoing testing. The baffle effect is observed mainly in the region where the probe is inserted. However, this perturbation is also found to reach the region below the impeller and to change the velocity profile there, resulting in differences in dissolution profiles when the tablets are fixed at positions that are downstream of the probe and within the low velocity region below the impeller.

On the other hand, the hydrodynamic effect generated by the probe does not appear to be particularly strong. In most dissolution testing runs, the changes in dissolution profile are not large enough to fail the tests, according to the FDA criteria (f_1 and f_2 values). The PIV measurements additionally show that the baffle effect is not strong enough to break the overall flow pattern, or to affect the region around the impeller, which is dominated by the main flow generated by the impeller.

It can be concluded that the hydrodynamic effects generated by the arch-shaped fiber optic probe are real and observable, resulting in slightly modification of the fluid flow in the dissolution vessel and therefore in detectable differences in the dissolution profiles. However, these effects are limited and do not typically lead to dissolution testing failures.

(371a) Towards the Combination of Modeling Strategies for Solid-Liquid Mixing

Blais, B., Ecole Polytechnique de Montreal
Bertrand, F., Ecole Polytechnique de Montreal
Originally presented on: 11/5/2013 15:15:00 - 15:33:00

Solid-liquid mixing is a topic that has not been completely explored due to the variety of challenges it faces. These challenges include the occurrence of complex flow patterns in the tank, which may be due to its geometry, to turbulence or rheological phenomena therein, as well as to the intrinsic multiscale nature of solid-liquid flows. In particular, it remains unclear how the interaction between turbulence or rheology and the particle collisions impact the flow patterns and critical mixing parameters such as the impeller torque, the just-suspension speed (N_{js}) and the dispersion characteristics. To shed light on these issues related to solid-liquid mixing, numerical and experimental work is essential.

The seminal paper of Tsuji [1] discusses and establishes that solid-fluid modeling strategies can be classified by the scale, macro, meso or micro, within which the fluid and the solid phases are modeled. This allows for a better understanding of the strengths and weaknesses of these methods, both in terms of precision and modeling hypotheses, as well as numerical challenges such as computational time. With these strengths and weaknesses in mind, more than one method can be integrated into a multiscale strategy. The development of a new drag correlation by Beetstra et al. [2] and its application to solid-liquid modeling is a prime example of the relevance of such an approach.

In the recent years, many numerical models have been used to simulate solid-fluid flows. A non-extensive list of these methods include the Eulerian-Eulerian [3] and Eulerian-Lagrangian approaches [4], the combination of CFD and the Discrete Element Method (CFD-DEM) [5], the Quadrature Method of Moment (QMOM) [6], the Multiphase Particle In Cell Methods (MP-PIC) [7, 8], Stokesian Dynamics [9] and DNS [10]. Despite their relative potential, many of these methods have not come of age and their use for the study of solid-liquid mixing flows is still rather limited.

In this work, we present the challenges linked with the physics of solid-liquid and its modeling. We classify the models available in the literature on the basis of their modeling scale. The principles behind each model are briefly explained in order to give a better understanding of their strengths, weaknesses and regime of applicability in terms of particle concentration and length scale. The potential combination of these models into multiscale modeling strategies is discussed.

This work is the first step towards the development of new multiscale strategies for the efficient modeling of solid-liquid mixing, with the aim of gaining a better understanding of the governing phenomena and underlying physics.

References

[1] Y. Tsuji, Multi-scale modeling of dense phase gas-particle flow, Chem Eng Sci, 62 (2007) 3410-3418.

[2] R. Beetstra, M.A. van der Hoef, J.A.M. Kuipers, Drag force of intermediate Reynolds number flow past mono- and bidisperse arrays of spheres, Aiche J, 53 (2007) 489-501.

[3] D. Guha, P.A. Ramachandran, M.P. Dudukovic, J.J. Derksen, Evaluation of large eddy simulation and Euler-Euler CFD models for solids flow dynamics in a stirred tank reactor, Aiche J, 54 (2008) 766-778.

[4] T. Srinivasa, S. Jayanti, An Eulerian/Lagrangian study of solid suspension in stirred tanks, Aiche J, 53 (2007) 2461-2469.

[5] Z.Y. Zhou, S.B. Kuang, K.W. Chu, A.B. Yu, Discrete particle simulation of particle-fluid flow: model formulations and their applicability, J Fluid Mech, 661 (2010) 482-510.

[6] R.O. Fox, P. Vedula, Quadrature-Based Moment Model for Moderately Dense Polydisperse Gas-Particle Flows, Industrial & Engineering Chemistry Research, 49 (2010) 5174-5187.

[7] D.M. Snider, An Incompressible Three-Dimensional Multiphase Particle-in-Cell Model for Dense Particle Flows, J Comput Phys, 170 (2001) 523-549.

[8] P.J. O'Rourke, D.M. Snider, An improved collision damping time for MP-PIC calculations of dense particle flows with applications to polydisperse sedimenting beds and colliding particle jets, Chem Eng Sci, 65 (2010) 6014-6028.

[9] J. Brady, G. Bossis, Stokesian Dynamics, Annual Reviews in Fluid Mechanics, 20 (1988) 111-157.

[10] J.J. Derksen, Direct numerical simulations of aggregation of monosized spherical particles in homogeneous isotropic turbulence, Aiche J, 58 (2012) 2589-2600.

(371b) CFD Modeling of Viscoelastic Droplet Breakup

Dietsche, L. J., The Dow Chemical Company
Parmar, N., The Dow Chemical Company
Originally presented on: 11/5/2013 15:33:00 - 15:51:00

When polymers having different rheologies are mixed together, there is often a desire to break up large polymer droplets into smaller ones, creating a more homogeneous blend with more desirable physical properties. It would be beneficial to have a model to help understand and predict how rheological characteristics will affect droplet breakup and blending. Building an empirical model requires the generation of a large number of polymer samples, followed by complex blending experiments in the melt phase. Another approach, which minimizes the need for extensive experimentation, is to utilize computational fluid dynamics (CFD) to explore the effects of the polymer rheology, equipment design, and operating conditions on droplet break-up efficiency. This is challenging since most polymeric materials are viscoelastic; in the melt state, they exhibit memory or stress relaxation responses and normal stress effects. In addition, a multiphase code is required to explore droplet deformation and breakup phenomena. Unfortunately, there are presently no commercial CFD codes that can handle both viscoelasticity and multiphase phenomena. This need prompted the development of a model using OpenFOAM® (open source CFD code) to simulate two-phase flow of viscoelastic materials. The numerical algorithm has been validated using two well-known viscoelastic free-surface effects, namely the die swell effect and the Weissenberg (rod climbing) effect. Additional effects such as transient and steady-state deformation of single drops subject to shear and planar elongational flows were also used for validation.

In this paper, we will examine some of the results obtained when using the OpenFOAM model to simulate viscoelastic droplet deformation and break-up characteristics in pure shear and elongational flow fields. The effects of viscosity and elasticity will be discussed. Additionally, the ability to simulate the observed phenomenon of droplet transverse stretching in a shear field will be discussed.

(371c) Hybrid Multiphase CFD Simulation for Interfacial Area Prediction in Liquid-Liquid Extraction
Wardle, K. E., Argonne National Laboratory
Originally presented on: 11/5/2013 15:51:00 - 16:09:00

A hybrid multiphase CFD solver which combines the Euler-Euler multi-fluid method with VOF sharp interface capturing has been developed using the open-source package OpenFOAM® for application to liquid-liquid extraction devices such as annular centrifugal contactors. This solver (*multiphaseEulerFoam*) has been extended to enable prediction of mean droplet size and liquid–liquid interfacial area through a single moment population balance method. Simulations of liquid–liquid mixing and separation in a model annular centrifugal contactor are reported. We compare several common droplet breakup/coalescence models along with efforts to calibrate model parameters versus available experimental data. Quantitative comparison is made for two different housing vane geometries and it is found that the predicted droplet size is significantly smaller for vane geometries which result in higher annular liquid holdup. Comparison with experimental observations for similar systems will be discussed along with proposed efforts to extend the methodology to include interfacial mass transfer and enable prediction of overall stage extraction efficiency.

(371d) Determination of Drop Size Distribution in Turbulently Agitated Liquid-Liquid Dispersion Using Numerical Simulations

Komrakova, A. E., University of Alberta
Eskin, D., Schlumberger
Derksen, J., University of Alberta
Originally presented on: 11/5/2013 16:09:00 - 16:27:00

Liquid-liquid dispersion and emulsion properties, such as apparent viscosity, rheology, stability, interfacial area available for heat and mass transfer, are determined by the size of droplets in the system. Accurate prediction and control of the drop size distribution (DSD) can be vital for practical applications.

In the present study, direct numerical simulations were utilized to determine the DSD of a turbulently agitated liquid-liquid dispersion. Large parallel computations were performed in a three-dimensional, fully-periodic domain using a free energy lattice Boltzmann equation model. The utilized numerical method is a diffuse interface method. No interface treatments or reconstructions are required. The interface evolves naturally due to thermodynamic mechanism employed. This issue allows to simulate systems with hundreds of drops during a reasonable time and affordable computational efforts. However, diffuse interface methods suffer from dissolution of small droplets. Our previous studies outlined that in order to mitigate the drop dissolution effect it is necessary to have the final size of the drops in the dispersed systems of 20-30 lattice units [lu]. Then the rate of dissolution significantly slows down, which is demonstrated in the present simulations.

Two liquids of equal viscosity $\mu=10^{-3}$; Pa·s and density $\rho=1000$ kg/m³ were considered. The dispersed phase volume fraction varied from 0.005 to 0.1. The simulations were carried out as follows: first, one-phase homogeneous isotropic turbulence was generated throughout the domain by means of linear forcing. Then the second phase was instantaneously injected. The energy input into the system varied to obtain Kolmogorov length scales of $\eta_k=1$, 5 and 10 lattice units [lu] in a fixed simulation domain of 500^3 size. The entire process of dispersion formation was visualized revealing the drop/eddy interactions on different scales. When equilibrium between drop breakup and coalescence was achieved the DSD was determined. As an example, the dispersed phase field at a Kolmogorov length scale of one lattice unit and the dispersed phase volume fraction of 0.07 is shown in Fig.1.

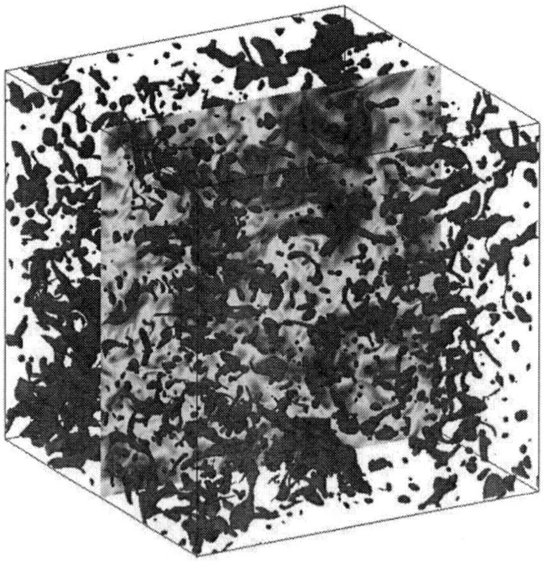

Fig.1. The dispersed phase field and the velocity magnitude in a cross-section at $\eta_k=1$ [lu] and the dispersed phase volume fraction of 0.07.

(371e) CFD Simulation of Integrated Physical and Biological Processes for Anaerobic Digestion

Wu, B., Philadelphia Mixing Solutions Ltd

Originally presented on: 11/5/2013 16:27:00 - 16:45:00

An extensive investigation of anaerobic methane fermentation requires identifying the relationship between physical environment and biological process. A computational fluid dynamics (CFD) based mathematical model was developed to characterize mixing and heat transfer coupled with biochemical reaction kinetics for anaerobic methane fermentation. The transient biological process was solved based on steady physical process, in which a computational cell was physically treated as an individual bioreactor with its own residence time and temperature. The CFD method was validated against two experiments available from the literature. An examination of mixing in a complete-mix digester indicates that the methane production remains almost unchanged while the energy efficiency decreases with increasing mixing power. A quantitative analysis of heat transfer in the digestion process of a plug-flow digester demonstrates that the energy output increases nonlinearly with the increase in heating energy within a limited temperature range. The CFD simulation platform is generic and may be applicable to the study of biohydrogen fermentation and other bioenergy sources.

(371f) Application of a Combination of Statistical Techniques and Mechanistic Modeling to the Mixing Process in Co-Rotating Twin Screw Extruders

Amedu, O. K., Northumbria University

Originally presented on: 11/5/2013 16:45:00 - 17:03:00

This paper proposes and experimentally validates theoretical models for the mixing process in a plasticating co-rotating twin screw extruders using a combination of statistical techniques and mechanistic modelling. A predictive model for the global residence time distribution was developed using a continuous exponential distribution model by examining the local contributions. Parameters such as the mean residence time, mean time delay, degree of fill, local drag and volumetric flow rates were obtained using mechanistic modelling by mathematically modelling the flow dynamics in the extruder. Predictive models of the mean residence time and time delay were subsequently developed. The effect of input conditions such as feed rate and screw speed on the shape of the residence time distribution curve and the degree of mixing were also evaluated by analysing the variance in this distribution.

Increase in feed rate at constant screw speed and vice versa results in narrowing the shape of the residence time distribution curve, with the feed rate having a higher sensitivity. The mean residence time and time delay was observed to drop with an increase in the feed rate. With an increase in the screw speed at constant feed rate, a reduction in the mean residence time was observed at the expense of an increased time delay. This has been validated against the experimental results.

(371g) Computational Study of Solids Mixing in Flotation Cells Using CFD-DEM Simulations

Basavarajappa, M., University of Utah
Miskovic, S., University of Utah
Originally presented on: 11/5/2013 17:03:00 - 17:21:00

In the present work, computational fluid dynamics (CFD) and discrete element method (DEM) have been coupled to study transport and mixing of discrete solid phase (quartz particles) in flotation cells. The off bottom suspension and subsequent homogenization of solid particles in the bulk fluid crucially depends on the impeller-stator design, impeller speed, off bottom clearance, and physical properties of particles. The turbulence level and distribution of turbulent kinetic energy (k) determines the extent of solid phase mixing and level of homogenization in the flow. At relatively high solids loading, inter-particle collisions become important and have to be modeled to accurately describe the flow. In the CFD-DEM approach, fluid motion is modeled using CFD, and solid particle motion and their interactions are simulated by DEM. During the coupling process, particle locations and velocities are updated in the CFD solver after a small time step and forces on the particles due to fluid motion are calculated. From the CFD solver, the forces on the particles are used in DEM to calculate new particle locations and velocities. The turbulence in the fluid phase is modeled by using Reynolds averaged Navier Stokes (RANS) based k-ε model. The contact model used in DEM is non-linear Hertz-Mindlin model – a soft sphere approach.

Particle size of 500 and 2000 μm at low solids loading (< 5%) are considered and the resulting solid flow dynamics are investigated. Modulation of turbulence due to preferential concentration in the case of smaller particles, and generation of wakes and their interaction for larger particles are also investigated. Furthermore, the turbulent zone close to the impeller in the presence of particles is fully characterized. Finally, the importance of stator in dispersing solid particles is additionally studied. This work represents a first attempt to model complex liquid-particle flows inside of flotation cells by using CFD-DEM method.

(371h) Erosion of Granular Beds - Solids Suspension Mechanisms

Derksen, J., University of Alberta

Originally presented on: 11/5/2013 17:21:00 - 17:39:00

Erosion of granular beds by a shear flow of liquid over its surface has many natural and engineered manifestations. In terms of industrial applications one can think of flow in slurry pipelines, dredging, and solids suspension in mixing tanks. The extent to which solids are entrained by the flow is a result of a competition between hydrodynamic stress and net gravity. The dimensionless Shields number embodies this competition and a critical Shields number demarcates between static and mobilized solids in the (top layers of the) granular bed.

Simulations of erosion processes are also interesting for fundamental and numerical reasons. The suspension flow closely above the bed is very inhomogeneous with a solids loading that can change from random close packing to dilute over a distance of only a few particle diameters. This poses challenges to continuum-based and point-particle-based numerical approaches; e.g. not much is known about the drag force in the presence of a strong solids volume fraction gradient; not every method can deal with suspensions having such wide ranges of solids volume fractions. Simulations in which we resolve the flow around the particles (?particle-resolved simulations') should be able to overcome the challenges but at the same time face issues of themselves. In addition to their stiff computational demand, these issues mainly relate to interactions between the particles when their surfaces are very close to or actually in (dry) contact. Then these ?direct' simulations need to revert to modelling (lubrication forces for near-contact and friction and restitution coefficients upon contact).

In this contribution we present different types of erosion-related flow systems that we have approached through resolved-particle simulations (with the particles being solid spheres). (1) Erosion of a granular bed by a laminar simple shear flow; (2) erosion of a granular bed by a mildly turbulent channel flow; (3) erosion of a solids bottom layer by the (again mildly turbulent) flow driven by an impeller in a mixing tank. Impressions of these flow systems are given in the figure.

The aim of this work is to enhance our understanding of the way solids are entrained by the liquid flow, and to assess how critical the overall results depend on numerical setings and on modelling choices regarding particle-particle (near) contacts.

(467a) Experimental Investigation of Multiphase Mixing in Annular Centrifugal Contactors Using High-Speed Imaging and Electrical Resistance Tomography (ERT)

Wardle, K. E., Argonne National Laboratory

Originally presented on: 11/6/2013 12:30:00 - 12:50:00

Advanced multiphase CFD methods are being developed and applied to simulation of centrifugal contactors to aid in optimization of their design and operation and provide inputs for design/simulation tools for processes based on this technology. Such complex multiphase simulations require detailed experimental data for validation of the underlying models and local predictions of dispersed phase droplet size and overall phase distributions (macro-mixing). *In situ* droplet size imaging and analysis has been conducted for evaluation of dispersed phase droplet size at several locations under the rotor of an engineering scale annular centrifugal contactor. The experimental setup employs a high-power, short-pulse strobe system in conjunction with a high resolution DSLR camera to obtain striking images of liquid droplets and air bubbles at typical operating conditions. In addition, large-scale spatial distribution of phases has been interrogated with electrical resistance tomography (ERT). Custom reconstruction techniques are employed to account for the complex geometry of the under-rotor region of the centrifugal contactor. Comparisons are made for different mixing vane configurations and corresponding predictions from CFD simulations to date.

(467b) Mixing Performance of a Bioreactor

Mattar, S.,
Machado, M. B., University of Alberta
Kresta, S., University of Alberta
Sauvageau, D., McGill University

Originally presented on: 11/6/2013 12:50:00 - 13:10:00

Bioreactors have been widely used in several research and industrial areas, such as the production of ethanol, but the fluid dynamics and the mixing performance of several geometries have not been determined so far. The flow field of an industrial scale bioreactor has been predicted using a CFD commercial package. The bioreactor is 12 m tall and has a diameter of 6.5 m. There are two tangential inlets and the outlet is located at the bottom of the vessel. The lower inlet has a flow rate of 300 kg/s and the upper inlet has a flow rate of 1200 kg/s. The software used for the simulation was Ansys CFX 14.0. A non-structured mesh was used. The turbulence models used were k-ε and Reynolds stress model. In order to assess the mixing performance of the vessel, some modifications were testes, such as: the addition of baffles and the variation of the position of the inlets.

(467c) Gas Dispersion With Up-Pumping Maxflo W Impellers

Shaik, M. U. R.,
Myers, K. J., University of Dayton
Janz, E. E., Chemineer, Inc.
Originally presented on: 11/6/2013 13:10:00 - 13:30:00

Despite known advantages of up-pumping wide-blade axial-flow impellers for gas dispersion, little is known about the design guidelines of these impellers. In the present study, the gas dispersion capabilities of the up-pumping axial-flow Maxflo W in low-viscosity liquids have been characterized, and compared with those of an up-pumping pitch-blade turbine and radial-flow CD-6. This comparison includes the speed, torque, and power requirements of the three impellers.

Testing at three scales (approximate volumes of 20, 70, and 170 liters) indicates that the dispersion capabilities of the Maxflo W can be described in terms of a scale-independent aeration-number Froude number relationship. Although a minimum impeller rotational speed is required to disperse low gas flows, the increase in dispersion speed is small with increasing gas flow rate. The effects of geometric parameters such as impeller to tank diameter ratio, sparger size, sparger location and liquid height on flooded to dispersed conditions have been investigated. Unlike radial-flow gas dispersion impellers that are not strongly affected by some geometric parameters, the up-pumping Maxflo W dispersion performance can be highly dependent on sparger size and location. Additionally, small impeller to tank diameter ratio Maxflo W impellers have been found to perform poorly in gas dispersion operations because of their highly axial discharge flow.

It was found that the up-pumping pitched-blade turbine exhibits time-dependent dispersion behavior that can permanently fail to disperse gas after appearing to be capable of dispersion for reasonably long periods of time. For this reason, collection of design data required extended, meticulous testing.

(467d) The Simulation of Solid Suspension By Using Scaba and A310

Hsu, R. C., Chang Gung University
Kang, J. C.,
Originally presented on: 11/6/2013 13:30:00 - 13:50:00

In this study, Computational Fluid Dynamics (CFD) modeling was developed for solid suspension in an agitation vessel. Two types of impellers were used: Scaba and A310. To validate the model, the simulated results for particle distribution at the vessel bottom are compared with the visualization experiments. The simulation results are in reasonable agreement with the experimental data, which suggest that the present CFD model can capture the essential features in solid-liquid flow systems. Owing to the gathering of the particles in the vessel center caused by the radial flow, the particle distribution shows a crisscross pattern for radial type impeller (i.e., Scaba impellers). The particle distribution in the vessel agitated by A310 was more homogeneous than that of the others due to both of the axial and downward flow nature of A310. Furthermore, modified correlations based on the Zwietering equation were obtained to account for the effect of impeller clearance from the experiments.

(467e) Improved Mixing and Mass Transfer Scale-Up: Design & Characterization for Cell Culture in Single Use Bioreactors

Jones, N., Thermo Fisher Scientific
Brau, C.,
Originally presented on: 11/6/2013 13:50:00 - 14:10:00

Improved Mixing and Mass Transfer Scale-Up: Design & Characterization for Cell Culture in Single Use Bioreactors

An in-depth study of common direct sparge mechanism in stirred tank bioreactors revealed that increasing gas entrance kinetic energy and vessel liquid column height were variables that must be characterized in order improve sparge performance and create a more predictable system. Improving the sparge mechanism sufficient to generate a targeted ratio of oxygen to carbon dioxide mass transfer ratio improves process development knowledge and allows for much lower variability when preparing for large scale processing required for biomolecule manufacturing. The precision drilled hole sparge has been carefully designed to generate a relatively constant bubble size by providing enough pores to avoid "pore saturation" at normal flow rates (0.001 to 0.1VVM). This design addresses the needs of scalable O2 kLa and CO2 stripping when performing cell culture at dissolved oxygen (DO) set-points of 30-50% airs saturation, while avoiding high velocity jetting of gas into the culture, reduces micro bubble formation. This allows more consistent performance, better scaling, and reduced foam generation.

Numerous studies have shown that gas entrance velocities near and in excess of 30m/s are harmful to shear sensitive cultures. The new sparge design delivers velocities at less than half of this threshold across all designs at 0.1vvm.

Supporting data will include a design of experiments study. These methods include characterization of bioreactors from 50L to 2000L in single use disposable formats. Critical to this research were high speed camera analysis, a novel method for measuring CO2 stripping, and gauge R&R type studies that help to identify sources of experimental error.

The culmination of this study includes cell culture experiments growing Chinese hamster cells to produce recombinant protein at 50L, 250L, and 2000L scale.

(467f) Foamless and Bubble Free Dynamic Gas-Liquid Mixing in Cell Culture Bioreactor

Cheng, A.,
Zhou, Y.,
Originally presented on: 11/6/2013 14:10:00 - 14:30:00

Advances in metabolic engineering have increased cell densities and yields in production of monoclonal antibodies and therapeutic proteins. However, scale-up become more difficult for high cell density cell culture mainly due to foaming, excessive dissolved CO_2 and uncontrolled osmolality levels. Increase in hydrostatic liquid head of larger scale bioreactors favors gas dissolution but not dissolved CO_2 removal. Attempts to remove excessive CO_2 by sparging with air are limited by the amount of foaming this cause. Countering excess foaming with antifoaming agents reduces gas exchange rate and fouls downstream processing equipment. Yield reduction up to 65% has been reported during scale-up to 10,000 L bioreactors. At present, no existing commercial process is capable of achieving these optimal levels without creating excessive shear and foaming. Against conventional wisdom, a dynamic surface gas exchange technique has been demonstrated to overcome a reduction in surface-to-liquid volume ratio during scale up. By controlling the dissolved CO_2 levels, sodium bicarbonate addition for pH control can be minimized, resulting in improved cell viability and product yield. As bioreactors are scaled up, the static liquid surface to volume ratio will decrease. However, it was discovered that dissolved CO_2 can be removed from large bioreactors without the typical process limitations. Data from over 100 fed-batch runs and 150 L pilot data will be summarized and presented.

(467g) Vortex Generation With Submerged Baffles: Comparison Of Two Scales

Giacomelli, J. J., Philadelphia Mixing Solutions Ltd.
Wyczalkowski, W., Philadelphia Mixing Solutions
Originally presented on: 11/6/2013 14:30:00 - 14:50:00

It is often desired in solid-liquid or gas-liquid applications of a stirred tank to use an upper located impeller and modified baffles to generate a stable vortex at the air-liquid interface to drawdown, wet, and incorporate floating solids or induce and disperse gas. The vortex for such applications must be stable, and for significant gas incorporation, must be of the minimum depth of the upper impeller submergence. To generate a centrally located stable vortex, partial baffles of standard width (T/10) can be terminated sub-surface at the location of the impeller. This produces rotational flow in the upper region of the vessel allowing a stable vortex to form. This configuration was tested in two scales, a 0.61m and 2m diameter tank, using Hydrofoils, Pitched Blade Turbines, and Six-Bladed Disc Turbine in impeller to tank diameter ratios of 0.3, 0.4, and 0.5. The geometric factors of interest were the impeller types, D/T and baffle/impeller-liquid coverage (IBC). The comparison and correlation between the two scales will be discussed along with scale up strategies and process implications.

(476a) Computational Fluid Dynamic Simulations of Gas-Liquid and Liquid-Liquid Flow in An Advanced-Flow Reactor (AFR)

Nieves-Remacha, M. J., Massachusetts Institute of Technology

Originally presented on: 11/6/2013 12:30:00 - 12:50:00

Computational Fluid Dynamic Simulations of Gas-Liquid and Liquid-Liquid Flow in an Advanced-Flow Reactor (AFR)

María José Nieves Remacha[1] and Klavs F. Jensen[1,*]

[1] Dept. of Chemical Engineering, Massachusetts Institute of Technology, Cambridge, MA 02139, USA

Microreactor technology has been demonstrated to be very useful in the laboratory scale to perform kinetic studies, optimization of reaction conditions, and catalyst screening with reduced waste generation, high mass and heat transfer rates, and automated operation, among other benefits. However, microreactors do not provide sufficient throughput for production purposes. Thus, the alternative is to have an intermediate-scale device that still holds microreactor technology advantages (such as mass and heat transfer performance) and can be parallelized with a reasonable number of units to achieve the production rate according to the demand.

One alternative is the Advanced-Flow Reactor (AFR) manufactured by Corning Inc. The mixing module is devoted for multiphase reactions and has a special design with several rows of heart-shaped cells in series that produce large specific interfacial areas (1,000 ? 10,000 m^{-1}) and overall mass transfer coefficients (0.1 ? 10 s^{-1}).[1]

The hydrodynamic characteristics of the AFR have been studied experimentally for carbon dioxide-water and hexane-water. Experiments can be tedious, expensive, and consume materials. Modeling allows us to study the effect of fluid properties and operating conditions (temperature, pressure, flow rates) on bubble/drop size distribution, specific interfacial areas, phase holdup, and mass transfer coefficients without the need of experiments. Computational fluid dynamic (CFD) simulations help us understanding the local behavior of the flow in the reactor, propose new designs, and select the best operating conditions and materials based on the CFD results.

Here we present results of CFD simulations using the open source software OpenFOAM® for steady-state single-phase (Figure 1 shows an example of the velocity profile obtained in the AFR) and transient two-phase flow (gas-liquid and liquid-liquid). For two-phase simulations, the interface capturing volume-of-fluid (VOF) approach was used, which corresponds to the *interFoam* solver included in OpenFOAM ®. Slight modifications to the code were performed in order to improve the results and these are also presented here.

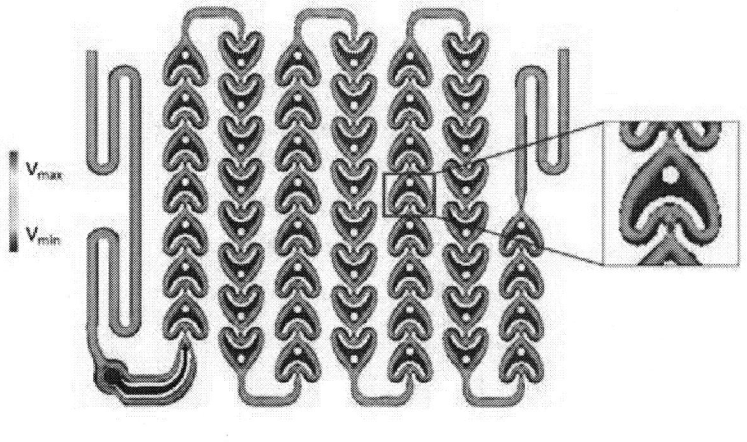

Figure 1: Velocity profile of CFD simulation of single-phase flow through the AFR using OpenFOAM®.

A first validation of the VOF method in simple geometries was performed to determine variables that are critical to guarantee a successful prediction of the biphasic flow characteristics. Some examples such as the prediction of two-phase flow in a T-junction are included in Figure 2 [2]. Results simulated with OpenFOAM® are compared with experiments and simulations with Fluent available in the literature [2-4].

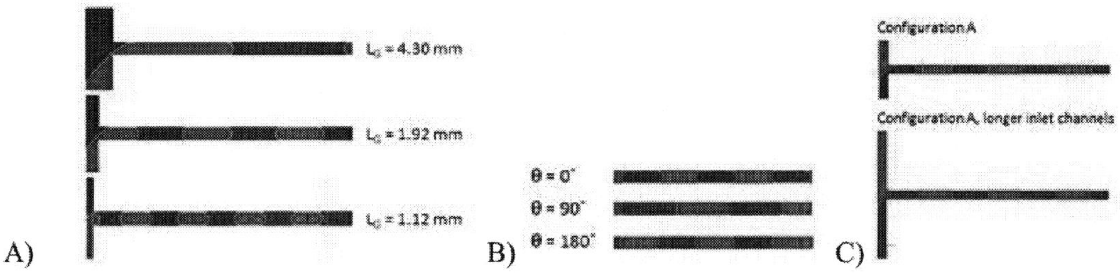

Figure 2: CFD Simulation of two-phase flow in a T-junction using OpenFOAM ®: A) Effect of inlet channel width on slug length; B) Effect of contact angle on flow characteristics; C) Effect of inlet channel length on slug size.

CFD simulations for hexane-water and carbon dioxide-water in the AFR are included in Figures 3 and 4 for different combinations of flow rates and orientations of the AFR. Fair agreement between simulation results and experiments based on the images of the flow obtained by high speed imaging was observed. Greater discrepancies were encountered when including gravity in simulations for liquid/liquid flow (as shown in Figures 3A) and 3C)), and better agreement between experiments and simulations for gas/liquid flow was observed (Figures 4A) and B)). Current analysis of the CFD results include measurement of bubble/drop size distribution, specific surface area, dispersed phase holdup, and mass transfer coefficients, and comparison with the experimental values. Future CFD simulations will include different fluid properties to study their effect on the hydrodynamic characteristics of biphasic flows. The current *interFoam* solver only solves the transient dynamics of the biphasic flow without mass transfer of components between phases or reaction terms. The ultimate goal is to include mass transfer and reaction so that prediction of conversion and selectivity for specific multiphase reactions can be performed.

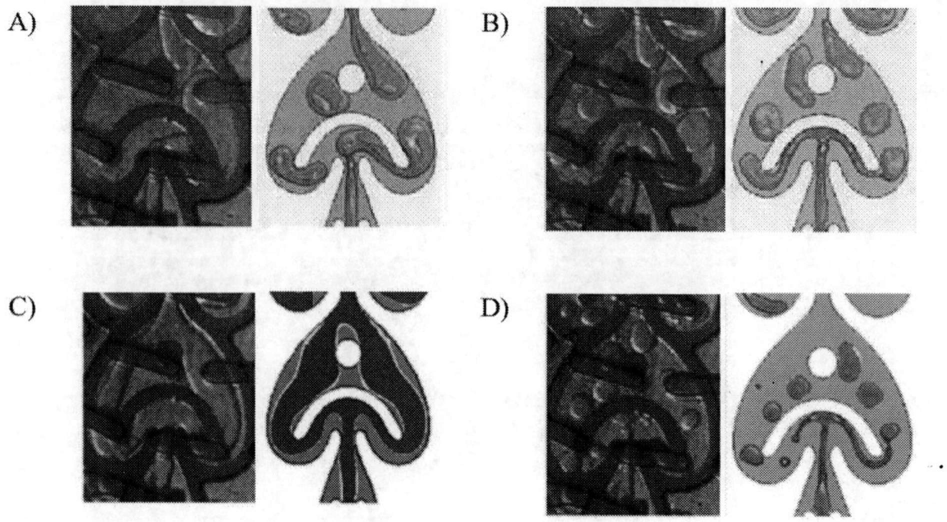

Figure 3: CFD Simulation of hexane/water flow using VOF in OpenFOAM® for vertical orientation of the AFR:

A) Q_w = 10 ml/min, Q_h = 10 ml/min; B) Q_w = 20 ml/min, Q_h = 20 ml/min; C) Q_w = 10 ml/min, Q_h = 40 ml/min; D) Q_w = 40 ml/min, Q_h = 10 ml/min

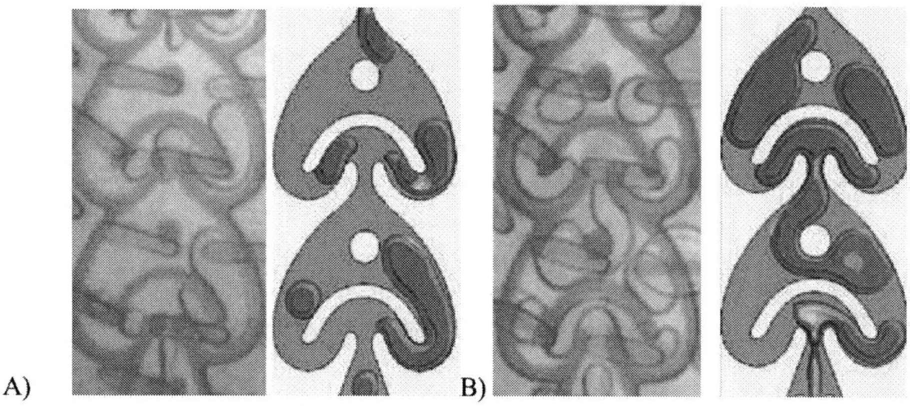

Figure 4: CFD Simulation of carbon dioxide/water flow using VOF in OpenFOAM® for Q_L = 10 ml/min: A) Vertical orientation, Q_g = 13.3 ml/min B) Horizontal orientation, Q_g = 20.8 ml/min

References:

[1] Nieves-Remacha, M. J.; Kulkarni, A. A.; Jensen, K. F. Hydrodynamics of Liquid-Liquid Dispersion in an Advanced-Flow Reactor. *Ind. Eng. Chem. Res.*, *51* (50), **2012**, pp 16251?16262

[2] Qian, D; Lawal, A. Numerical study on gas and liquid slugs for Taylor flow in a T-junction microchannel. *Chem. Eng. Sci. 61 (23)* **2006** pp 7609?7625

[3] Santos, R.; Kawaji, M. Numerical modeling and experimental investigation of gas?liquid slug formation in a microchannel T-junction. *International Journal of Multiphase Flow 36, 4* **2010** pp 314-323

[4] Vandu, C.O.; Liu, H; Krishna, R. Mass transfer from Taylor bubbles rising in single capillaries. *Chem. Eng. Sci., 60* **2005** pp 6430 ? 6437

(476b) Uncertainty Quantification in Coal Gasifier Simulations For Reliable Predictive Computational Fluid Dynamics Models

Gel, A., Alpemi Consulting LLC
Shahnam, M., National Energy technology Laboratory
Subramaniyan, A. K.,
Originally presented on: 11/6/2013 12:50:00 - 13:10:00

Although global coal consumption rose only by 3.1% (a below average) in 2008, coal remained the fastest growing fuel in the world for the sixth consecutive year. About 50% of electricity in the United States is being generated by coal. Demand for electricity is projected to increase by 30% over current demand levels by 2030. In 2008, global emission of CO_2 due to fossil fuel was approximately 30 petagrams, of which the United States accounts for about 19 percent. Future CO_2 emission is likely to be higher, unless considerable changes to the energy system are made. The U.S. Department of Energy has great interest in technologies that will lead to reducing the CO_2 emission of fossil fuel burning power plants. Pure streams of CO_2 can be successfully captured and stored and therefore reduce the impact of CO_2 to our atmosphere. Advanced power generation plants such as Integrated Gasification Combined Cycle (IGCC) can potentially lead to capture and storage of CO_2. In an IGCC power plant, a gasifier is used to convert coal into synthesis gas (syngas). After removal of impurities and pollutants from the syngas, the resulting gas is used to generate electricity in a gas turbine.

Advanced modeling and simulation capabilities have the promise of significantly reducing the time and cost of the development and deployment of technological processes such as gasification technology. Simulation technology allows rapid scale up of technologies, reducing or even potentially avoiding costly intermediate scale testing. New designs can be tested via simulations to ensure reliable operation under a variety of operating conditions. However, to ensure their usefulness and adoption in practice, the credibility of the simulations needs to be established with uncertainty quantification and probabilistic validation methods. Implementation of uncertainty quantification techniques can also provide guidance for design optimization under uncertainty to achieve robust designs for commercial scale advanced coal gasifiers.

In this study, Computational Fluid Dynamics (CFD), is used to study the quality of syngas, defined as H2/CO and CH4/H2 ratios, in a fluidized bed gasifier under varying operating conditions. The simulation results will be compared probabilistically to experimental data, which is available for the same fluidized bed gasifier. In the current study, the effect of variation in critical operating parameters such as coal feed rate, particle size and steam to oxygen ratio on the syngas quality is investigated using a non-intrusive uncertainty quantification techniques including Bayesian analysis and polynomial-chaos expansions. Smolyak sparse grids and optimized Latin hypercube sampling techniques will be used along with adaptive Bayesian optimization to reduce the number of expensive simulations. These sampling techniques enable interrogating correlated random variables with minimum (optimal) samples.

Additionally the extensive experimental data, which is available for the fluidized bed gasifier in this study, allows quantification of model form of uncertainty through Bayesian uncertainty quantification methods. The resulting discrepancy model (a hybrid combination of simulations and experiments) will be used to verify the validity of the simulation models over the entire design space. The reduced order models will be fully stochastic, i.e., will predict the entire probability distributions of the outputs for every input configuration. The probabilistic nature of these models is inherently suited for modeling risk in addition to the quantities of interest.

These validated predictive models based on CFD models would play a crucial role in scale-up studies as decision support tools and reduce time-to-market deployment of the advanced clean coal energy technologies.

(476c) Experimental and CFD Study of Methane Steam Reforming in a Catalytic Plate Reactor and Characterization of the Coated Noble Metal Catalyst

Mundhwa, M., Queen's-RMC Fuel Cell Research Centre
Peppley, B. A., Queen's University
Thurgood, C. P., Royal Military College
Originally presented on: 11/6/2013 13:10:00 - 13:30:00

In recent years catalytic plate reactors/reformers (CPR) have received considerable attention due to their high thermal efficiency, low pressure drop and low mass transfer limitation. This work focuses on an experimental study of methane steam reforming in an experimental catalytic plate test reactor. Performance data collected at various operating temperatures, steam to carbon ratios and gas hourly space velocities will be presented. The catalyst being studied consists of a noble metal reforming catalyst applied to a high heat transfer alloy metal substrate. The catalyst coating is characterized for active metal surface area, total surface area, noble metal dispersion and active phase distribution which can be used in the validation of a mathematical model that couples heat transfer and kinetics and leads to a tool for the optimization in plate reactor design.

The computational study of a catalytic plate reformer is very much limited to simplified assumptions such as plug flow, binary diffusion, and one dimensional energy flow across catalyst layers. Also the source term for the mass transport *i.e,* reaction rate is based on power law and Langmuir-Hinshelwood-Hougen-Watson kinetic models. Such limited transport assumptions and kinetic models lead to an incomplete understanding of the reformer behavior under steady as well as unsteady state. Hence, a detailed integral two dimensional finite element model that accounts for mass, heat and fluid dynamics, along with detailed microkinetics is developed for optimal design and control of a CPR.

(476d) Three Dimensional CFD-Based Numerical Study of Chemically Reacting Carbon Particles Within a Randomly Packed Bed

Schulze, S., TU Freiberg
Nikrityuk, P. A., TU Freiberg
Meyer, B., TU Freiberg
Originally presented on: 11/6/2013 13:30:00 - 13:50:00

This work is devoted to three-dimensional numerical study of the influence of gas flow (consisting of O2, H2O and N2) and its composition on the carbon consumption evolution within a randomly-packed 3D packing of spherical non-porous carbon particles. The monodisperse packing was numerically generated using gravity-forced sedimentation which was modeled utilizing a in-house developed Discrete Element Method code.

The primary interest of this work is the spatial distribution of the temperature and species concentrations in the so-called combustion zone of a fixed bed, where the oxygen reacts with carbon particles. It should be noted that due to the inaccessibility to the interior of a fixed bed reactor experimental studies are not always capable of characterizing basic features of flow and phenomena occurring inside the reactor. From this point of view numerical simulations can play the role of 'numerical-experiments' enabling to see the processes 'in situ'.

In this work the reacting particles are represented by moisture and ash free nonporous carbon. The model includes six gaseous chemical species (O2, CO2, CO, H2O, H2, N2). Three heterogeneous reactions (C+O2, C+CO2 and C+H2O) and two homogeneous semi-global reactions, namely carbon monoxide oxidation and water-gas shift reaction, are employed. Several semi-global reaction rate expressions taken from the literature were utilized. The Navier-Stokes equations coupled with the energy and species conservation equations are used to solve the problem in pseudo-steady state approach. At the surface of the particle, the balance of mass, energy and species concentration is applied including the effect of the Stefan flow. The range of parameters considered in this study are $d = 2*10^{-2}$ m, Tin=1000 K, Re =100, where d, Tin and Re are the diameter of the particle, the inlet temperature of the gas and the Reynolds number, respectively. The inflow gas composition corresponds to dry air gas. The unstructured grid used in simulations comprises of approximately 6 millions of control volumes.

The results of numerical simulations were compared with those for a simple cubic packing of chemically reacting carbon particles. The comparative simulations showed that the penetration length of the flame decreases from four particle diameters for regular packing to one particle diameter in the case of random packing. Additionally, we found out, that the temperature and species gradients between particle surface and gas phase predicted for the random packing decreases, too. Spatial distribution of species, temperature and gas velocity are illustrated and discussed.

(476e) Development of a Methodology to Obtain the Kinetic Mechanism of Steam Methane Reforming Compact Reactor Using CFD Coupled With Hybrid Genetic/Surface Response Optimization Model

Moreira, R. P. M., ESSS
Almeida, L., ESSS
Bacchi, R. D. A., ESSS
Ropelato, K., ESSS
Passarelli, F. M.,
Quintella Coelho, E. F. A., Petrobras
Bidart, A. M. F.,
Originally presented on: 11/6/2013 13:50:00 - 14:10:00

The process intensification characteristics of compact reactors (enhance heat and mass transfer properties) represent a breakthrough technology for heat and mass controlled reactions since it allows exploring the full potential of catalytic systems during highly endothermic or exothermic reactions. Concerning this issue, special attention is drawn to the steam methane reforming (SMR) reactor, a widely-used process to produce hydrogen or syngas (mixture of hydrogen and carbon monoxide). A conventional steam methane reformer (SMR) consists of several hundred fixed-bed reactor tubes filled with supported catalyst particles, which converts a mixture of hydrocarbons and water into hydrogen and carbon monoxide via an endothermic reaction. The SMR reaction is supported by the heat provided by fuel combustion in the process furnace (exothermic reaction). Besides the large use of the conventional technology, efforts have been made to improve process efficiency, reduce production cost and footprint, mainly for small-scale applications (offshore methanol and GTL plants, fuel cells applications, car fuelling stations, etc). Steam methane reforming (SRM) using compact reactor has great potential to be a **cost**-effective means to achieve such goal. The reaction mechanism for the SMR and catalytic combustion reaction is a key step to optimize the compact reformer reactor. Despite many kinetic mechanisms for the catalytic combustion and steam methane reforming reaction available, concerning different number of reactions, type of catalyst, inhibition effects by carbon dioxide and water in literature, information available is sometimes controversial and cannot be used in all cases.

This article aims to show the application of CFD coupled to a genetic algorithm optimization in order to obtain a reaction mechanism for steam methane reforming, water gas shift and combustion inside a catalytic reactor. The reactions were simulated considering a catalytic compact mini-channel reactor composed by stacked plates containing parallel arrays of mini-channels. The process channels (steam methane reaction) are filled with precious metal catalyst and the necessary heat for the reaction is provided by the intercalated plates containing parallel arrays of catalytic combustion channels. The optimization methodology makes it possible for the engineering professional to prevent the tedious task of manually executing analyses in a try-and-error approach, introducing a methodology where numerical simulations are executed without any user intervention. Multi-Objective Genetic Algorithm is used in the minimization of error between numerical results and experimental data. Four objectives, based on the summation of the squared relative error in each point, were used and 21 kinetic parameters of reaction rate equations of single step methane reforming, water gas shift and methane combustion were evaluated as optimizing variables. Due to heat transfer effects between the plates, catalyst and the gas mixture, the standard techniques to obtain the reaction rate could not be applied. Instead of that, a CFD simulation of a representative domain was used to calculate the conjugate heat transfer together with reactive flow. The optimization was applied to minimize the relative error between a simplified CFD simulation, which evaluated a two-dimensional reactive flow and the experimental results that could be obtained on the real scale equipment. The methane conversion and temperature for the combustion and reforming ducts were evaluated. The reaction mechanism obtained will feed a more complex CFD study, considering the complete reactor geometry, allowing the prediction of hot spots and mal distribution in some regions. Also, the model makes it possible to analyze the thermodynamic equilibrium for coke formation and the methane conversion for several flow rates and gas composition, which allows provides a safety operational window for the process.

(476f) Numerical Investigation of Packed Bed Reactors With Non-Spherical Particles

Eppinger, T., Technische Universitat Berlin
Jurtz, N., Technische Universitat Berlin
Kraume, M., Technische Universitat Berlin
Originally presented on: 11/6/2013 14:10:00 - 14:30:00

Introduction

Packed bed reactors are widely used in the chemical and process industry, amongst others for highly exothermic or endothermic catalytic reactions. Such reactors are characterized by a small tube to particle diameter ratio (D/d-ratio) to ensure a safe thermal management. For the design of such apparatuses the well-known correlations for packed bed reactors can not be used, because these reactors are dominated by the influence of the confining wall, which affects the porosity and the velocity field and as a result also the species and temperature distribution within the bed.

Due to the increasing computational power and the possibility of a detailed analysis of the transport phenomena in such reactors CFD (computational fluid dynamics) becomes more and more an indispensable tool. In the last years several authors have investigated mass, energy and species transport in spatially resolved packed beds mostly consisting of spheres, although non-spherical particles are much more common in industry.

Aim of this contribution

The aim of this contribution is to show firstly a fully computational work flow for the generation of packings consisting of non-spherical particles, meshing and calculation of the flow, temperature and species field and secondly to compare these packings in terms of radial porosity, velocity and temperature distribution.

Methods

In our work we have investigated packings of more complex particle shapes like cylinders and Raschig rings, which are often used in industry. The whole numerical process includes the modeling of these particles, the generation of the packed bed with the discrete element method (DEM), the meshing of the complex packing geometry as well as the CFD calculation which is done with the commercial tool STAR-CCM+ by CD-adapco.

For the generation of the packed bed DEM is used. DEM is based on solving Newton's Law for spherical particles. More complex particles like cylinders or Raschig rings can be modeled by combining several spheres to approximate the geometrical shape.

For the meshing of the packed bed for the CFD calculation the approximated particles are replaced with the original exact shapes. The meshing difficulties in the vicinity of the contact points of two particles or between the wall and the particle is handled by a method which is already published by the authors and was presented at AIChE annual meeting 2012: If the distance between two surfaces fall below a predefined value, the surface is modified locally resulting in a small gap. This gap can then be filled with cells of a reasonable quality and does not significantly affect the results.

Results

With the described methods we have investigated three different particle shapes (cylinder, Raschig ring, and 4 hole cylinder) with a D/d-ratio between 5.0 and 10.0 and a particle Reynolds number Re_p= 5 - 5000. All packings consists of up to 1500 single particles which is equal to a packing height of approximately 20d. Heat transport is simulated by applying a constant wall temperature.

The resulting packings are validated in terms of global bed porosity, radial and axial porosity profile with experimental results from literature. The results agree very well with these data. Where possible the CFD calculations are validated against literature data, too. The results are in a reasonable agreement for pressure drop and the radial velocity distribution.

Further investigations show, that the well-known dependence between the radial porosity profile and the radial velocity profile is not valid for Raschig rings. Additionally, the radial velocity is higher for all investigated particle shapes with a maximum for the cylindrical shapes compared to a packing of spheres. Furthermore the higher radial velocity leads to a higher convective radial heat transport which is beneficial for the thermal management of the reactor.

(476g) Computational Fluid Dynamics for the Analysis and Optimization of a Lab-Scale Stirred Tank Reactor With Catalytic Basket

Fonte, C. P., Faculdade de Engenharia da Universidade do Porto

Braga, M., Laboratoire de Génie des Procédés Catalytiques, UMR 5285 CNRS, CPE, UCB Lyon

Brunet-Errard, L., IFP Energies Nouvelles

Pitault, I., Laboratoire de Génie des Procédés Catalytiques, UMR 5285 CNRS, CPE, UCB Lyon

Simoëns, S.,

Bellefon, C. D., LGPC - CNRS/CPE Lyon

Santos-Moreau, V., IFP Energies Nouvelles

Lopes, J. C. B., LSRE - Laboratory of Separation and Reaction Engineering - Faculty of Engineering - University of Porto

Originally presented on: 11/6/2013 14:30:00 - 14:50:00

Lab-scale stirred tank reactors with catalytic baskets are used in petroleum and biofuels R&D to perform kinetic studies and primary screening tests for the selection of shaped catalysts (beads and extrudates). Different reactive systems, liquid-solid and gas-liquid-solid, are studied in these kind of reactors: selective hydrogenations, hydroconversions, transesterifications, etc. With the increasing catalysts efficiency and the use of complex fluids like heavy crude feedstocks or biomass derived, tests and catalyst selection in lab-scale stirred tank reactors are often hindered by hydrodynamics and mass transport limitations. Achieving a kinetically controlled operational regime is particularly important for fast or consecutive-competitive reactions.

For the study and optimization, the present work concerns the development of a 3D Computational Fluid Dynamics (CFD) model of the gas-liquid-solid flow in catalytic basket reactors. A commercial lab-scale reactor with a radial six-bladed self-inducing hollow impeller and a fixed annular basket was selected for the study. Turbulence in the tank was simulated with the k-ε model and the MRF approach was chosen to deal the impeller motion. The flow in the basket filling was simulated as a homogeneous porous medium with the Brinkman-Forchheimer equations and the basket walls were simulated as semi-permeable zero-thickness surfaces with a pressure drop. The Eulerian model was chosen for the simulation of the gas-liquid multiphase flow. Numerical results were compared with experimental data from Particle Image Velocimetry and a bubble tracking imaging technique.

CFD simulations provided the velocity field and the gas volume fraction in each point of the domain for different operational conditions, which were used for the prediction of the mass transfer rates (k_S and k_La). The developed CFD flow model can then be used as an optimization tool for the improvement of reactor configuration or to set optimal operational conditions for kinetic and screening tests.

(578a) CFD Model of Suspension Polymerization in Mixed Reactors

Vonka, M.,
Soos, M., ETH Zurich
Morbidelli, M., ETH Zurich
Originally presented on: 11/6/2013 15:15:00 - 15:35:00

The properties of polymer produced by suspension polymerization (SP) are determined the particle size distribution (PSD) of dispersed polymer particles. We aimed our effort at filling the gap in description of the SP in industrial reactors.

SP process starts with dispersed droplets of monomer in the water phase. In case of bead SP, the reaction changes monomer droplets into a mixture of polymer and monomer. During the polymerization the droplet viscosity increases in orders of magnitude which strongly influences the rates of the agglomeration and breakage. In well-mixed experimental vessels the PSD is determined by particle breakage due to the homogeneity of the flow field. On the other hand, industrial reactors have large settling zones, where agglomeration of particles takes place.

Mixed vessel CFD model was coupled with population balance equations (PBE) of dispersed phase. The polymerization kinetics model provided the evolution of average polymer chain length and subsequently the viscosity of the dispersed phase. The simulations predicted PSD of the dispersed phase and it's dependency on the geometry of the vessel, mixing frequency and viscosity evolution during the polymerization. Since the polymerization and the fluid mixing have different time scales, a fully coupled CFD + PBE model describing the whole polymerization is computationally unfeasible. Steady state simulations were carried out following the viscosity evolution. Agglomeration and breakage kernels were adopted form literature [1, 2] and the closure problem of PBE was overcome by Quadrature Method of Moments (QMOM) [3].

This approach determines the significance of the variable viscosity on the PSD. Such an approach can be used for optimization of mixing conditions in large industrially used reactors.

Used literature:

1. Kotoulas, C. and C. Kiparissides, *A generalized population balance model for the prediction of particle size distribution in suspension polymerization reactors.* Chemical Engineering Science, 2006. **61**(2): p. 332-346.

2. Laakkonen, M., V. Alopaeus, and J. Aittamaa, *Validation of bubble breakage, coalescence and mass transfer models for gas-liquid dispersion in agitated vessel.* Chemical Engineering Science, 2006. **61**(1): p. 218-228.

3. Petitti, M., et al., *Bubble Size Distribution Modeling in Stirred Gas-Liquid Reactors with QMOM Augmented by a New Correction Algorithm.* Aiche Journal, 2010. **56**(1): p. 36-53.

(578b) Simulation of Mixing in Hot Melt Extruders Using Smoothed Particle Hydrodynamics

Eitzlmayr, A., Graz University of Technology

Radl, S., Graz University of Technology

Koscher, G., Research Center Pharmaceutical Engineering GmbH

Khinast, J. G., Graz University of Technology

Originally presented on: 11/6/2013 15:35:00 - 15:55:00

Introduction

Hot melt extrusion (HME) attracted increasing attention in pharmaceutical manufacturing during recent years due to its mixing capabilities for high-viscous materials and its continuous processing characteristics. The potential of HME to achieve solid solutions is a possible way to overcome the challenge of bioavailability of poorly soluble drug molecules, where the majority of newly developed drug molecules belongs to.

The most common devices used for pharmaceutical HME are co-rotating twin screw machines. Compared to single screws and counter-rotating twin screws the co-rotating twin screws are frequently preferred due to their mixing performance and self cleaning screw profiles. The typical modular screw design of co-rotating twin-screw extruders allows flexible use of a single device.

The HME process is used in the polymer industry for many years. Nevertheless, it is still a challenge to predict non-Newtonian, non-isothermal flow and mixing in the complex, rotating geometry of twin-screw extruders. This is especially the case for partially filled screw sections, for which conventional (mesh based) CFD simulations are currently not suitable[1]. To overcome these challenges, we follow a different strategy and investigate the applicability of the Smoothed Particle Hydrodynamics (SPH) method to the simulation of mixing in HME processes.

Method

SPH is a mesh-less Lagrangian particle method, which approximates the governing equations for fluid flow and can inherently handle free surface flows[2]. We use the open-source particle simulator LIGGGHTS (www.liggghts.com), which was originally developed for the simulation of granular flow via the Discrete Element Method (DEM), and provides an SPH module.

Different approaches to model wall boundaries in SPH exist, mostly based on particles. However, it is not obvious in SPH how to handle complex shaped geometries (as extruder screws), typically represented by a set of wall triangles. We developed a new approach to achieve an appropriate interaction between wall triangles and SPH particles.

Results

Figure 1 shows snapshots of the distribution of two different particle types (yellow and red) in a cross section of a co-rotating twin-screw extruder, calculated with our method. Figure 2 shows the top view of a similar extruder, and illustrates how a slice of tracer (shown in red) moves and disperses in axial direction due to the action of the screw. This allows a detailed analysis and comparison of the mixing performance of different screw geometries, e.g., conveying elements or kneading elements.

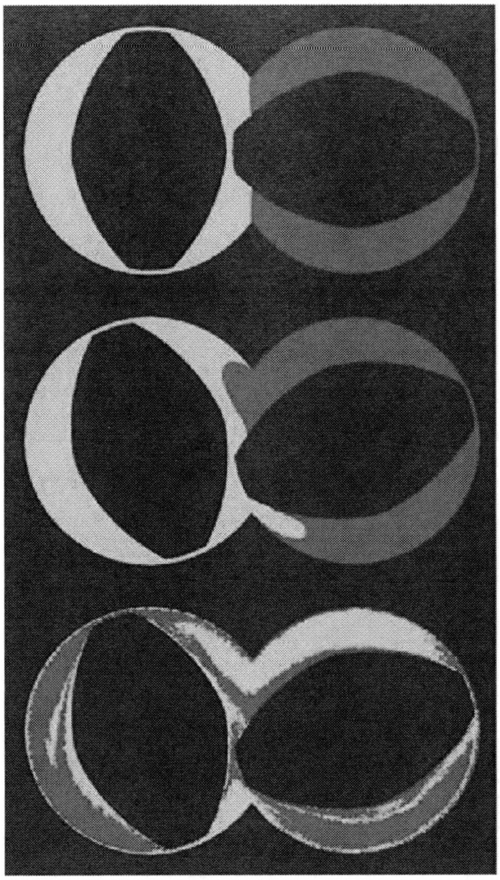

Figure 1: Snapshots of mixing in a co-rotating twin-screw extruder (cross section; top: initial state; center after 0.05 revolutions; bottom after 1.05 revolutions).

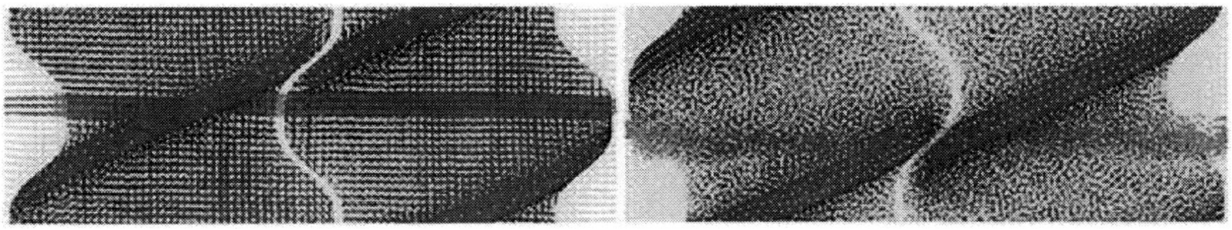

Figure 2: Axial mixing of a tracer in an extruder (top view; left: initial state; right: after 0.2 revolutions).

References

1. Sarhangi Fard A, Anderson PD. Simulation of distributive mixing inside mixing elements of co-rotating twin-screw extruders. *Computers & Fluids*. 2013. (in press)

2. Monaghan JJ. Smoothed particle hydrodynamics. *Reports on Progress in Physics*. 2005;68(8):1703-1759.

(578c) Investigation of Single and Multiphase Fluid Flow Behavior in Stirred Tanks By Means of CFD and Radioactive Particle Tracking

Bashiri, H., Department of Chemical Engineering, Ecole Polytechnique de Montreal, Quebec, Canada
Alizadeh, E., Ecole Polytechnique de Montreal
Bertrand, F., Ecole Polytechnique de Montreal
Chaouki, J., Ecole Polytechnique de Montreal
Heniche, M., Ecole Polytechnique de Montreal
Originally presented on: 11/6/2013 15:55:00 - 16:15:00

The understanding of the flow patterns and local hydrodynamics in stirred tanks is important for process design and scale-up. Thanks to the availability of faster and faster computers, the industry has shown in recent years an increasing interest in the use of CFD models for the simulation of single and multiphase fluid flows. However, the closure rules inherent to such complex models (e.g. for turbulence and phase interactions) often lead to uncertainties and further justify the need for experimental data obtained through reliable validation techniques. The opaque nature of the vessel walls or the fluids themselves renders inefficient conventional measurement techniques such as optical and laser-based methods (e.g. LDA, PIV). On the other hand, the so-called Radioactive Particle Tracking (RPT) technique is a suitable method that has been used to provide extensive data about liquid flow patterns for both single and multiphase flow systems. It is based on the detection of the Lagrangian motion of a radioactive tracer inside the vessel. Our research group has employed this technique for many years to provide local hydrodynamic data in various chemical reactors and mixing processes.

This work focuses on the assessment of CFD simulations of single and multiphase turbulent fluid flow in a laboratory-scale stirred tank using RPT data. First, the effect of impeller type, rotational speed and clearance on the flow patterns and agitation behavior of single-phase systems are investigated. Three types of impeller are considered: a Rushton turbine, a Pitched blade turbine (PBT) in down-pumping mode and a concave disc impeller (CD-6). Next, the mixing behavior of gas-liquid flow in the loaded and completely dispersed regimes with the aforementioned impeller types are investigated numerically with a Eulerian-Eulerian approach and experimentally using RPT. Finally, the importance of developing multiscale models for such complex flows is discussed.

(578d) Computational Fluid Dynamics Investigation of Mechanical Mixing in Anaerobic Bioreactors

Wang, J. J., Zhejiang University
Ge, C. Y., Zhejiang University
Gu, X. P., Zhejiang University
Feng, L. F., Zhejiang University
Wu, B., Philadelphia Mixing Solutions Ltd
Originally presented on: 11/6/2013 16:15:00 - 16:35:00

Methane from anaerobic digestion of organic matter is a source of biorenewable energy. Mixing in bioreactors helps maintain uniform total solids (TS) distribution and maximizes the direct contact between active biomass and raw sludge. This study presents a computational fluid dynamic (CFD) technique that evaluates the mechanical mixing performance quantified by blending time with respect to power input in bioreactors. In the model development, the digestate was assumed to be a non-Newtonian power-law fluid that is dependent on the TS level, and mixing was achieved by a standard pitched blade turbine (PBT) or a modified PBT (m-PBT) impeller. The mixing experiments were conducted using xanthan gum, in which the axial and radial velocities obtained with particle image velocimetry measurements were used to examine the CFD results. The power and flow numbers of an impeller were verified with the *laboratory* specifications. A general procedure for each run includes two stages: (1) using a multiple reference frame approach to predict steady-state turbulent mixing, and (2) solving a transient scalar transport equation to obtain the blending time based on the converged flow fields. An optimum mixing design was proposed based on the lowest power input required to homogenize the substrate in the bioreactors.

(578e) Validation of Openfoam Solver for Mixing Tank Analysis

Kasat, G., Tridiagonal Solutions Inc.
Vedapuri, D., Tridiagonal Solutions Inc.
Hule, P., Tridiagonal Solutions Inc
Bapat, Y.,
Originally presented on: 11/6/2013 16:35:00 - 16:55:00

Open source CFD tools such as OpenFOAM has gained significant attention and interest in the Industry. OpenFOAM has an extensive range of features to solve problems ranging from complex fluid flows involving chemical reactions, turbulence and heat transfer, to solid dynamics and electromagnetics. In-spite of all these capabilities its commercial use is limited because of non-availability of industry standard benchmarks. In this study, we systematically investigate the solver quality of the open source CFD code OpenFOAM for computation of the single phase as well as multiphase flow in stirred tank reactors. This study aims at discussing various case studies where OpenFOAM was successfully employed for predicting performance of stirred reactors in terms of power consumption, flow pattern, mixing time etc. The results predicted by OpenFOAM are compared with experimental data and commercial CFD tool Fluent (of ANSYS Inc.). The OpenFOAM solver was found to capture the single phase as well as multiphase flow in stirred tank with reasonable accuracy. The following case studies are discussed in this presentation.

Case – 1: In the first case the turbulent flow generated by Rushton Turbine impeller was solved using both OpenFOAM and Fluent. A mesh having combination of hexahedral and tetrahedral elements generated using Gambit (of ANSYS Inc.) was used for both the solvers. Steady state simulation was performed using MRF approach. The predicted velocities were compared with the experimentally measured velocities at various locations in the reactor.

Case – 2: In the second case study, capability of OpenFOAM solver was tested to capture the transient flow in stirred tank was using sliding mesh model. For OpenFOAM simulation a mesh generated by snappyHex mesh was used. The results predicted were compared with those predicted by Fluent using the mesh generated by Gambit.

Case – 3: In this case study, laminar non-Newtonian fluid in stirred reactor was solved using OpenFOAM and Fluent. The results were compared with respect to flow pattern, forces on impellers and baffles etc.

Case – 4: This case study focuses on testing capability of OpenFOAM solver to capture the gas entrainment in partially baffled stirred reactor. Volume of Fluid (VOF) model along with sliding mesh methodology was used to capture the entrapment of gas from headspace through the vortex. The results of OpenFOAM simulation (with snappyHex mesh) were compared with that of Fluent (using Gambit mesh).

(578f) Steady State CFD for Determining Just Suspended Speed

Russ, D. C., University of Louisville
Berson, R. E., University of Louisville
Originally presented on: 11/6/2013 16:55:00 - 17:15:00

A new method for determining the just suspended speed (Njss) in stirred tank reactors was developed using steady state CFD simulations. This method can be used for determining Njss for those systems in which it would be difficult with traditional experimental techniques. It also removes subjectivity associated with traditional experimental visualization techniques. Additionally, since the technique employs steady state simulations, it is advantageous over time-dependent simulations, which require much longer computational times due to the multiple phases and transient nature. Using an application of mean age theory, the method solves scalar transport equations within the settling zone along the tank bottom. Three-dimensional simulations have been performed for a range of solids concentrations and various tank geometries. Computational results compare very favorably with published experimental data. Plotting average age for the solid phase in this region versus impeller speed reveals three distinct regions, with the narrow central region corresponding to Njss. The three regions are believed to be caused by the interaction between solid fraction in the region and z-direction velocity of the solid phase: above Njss, the average age is low, due to large z-velocities; below Njss, the average age is also low, due to the higher concentration of the solid phase allowing for greater scalar transport; at Njss, the mean age spikes sharply due to the lack of either of these two factors.

(578g) Discrete Tangent Based Sensitivity Analysis Method in Computational Fluid Dynamics (CFD) for Stirred Tank Reactors

Srinivasan, V., Amoeba Technologies, Inc.
Sathyamurthy, P., Amoeba Technologies, Inc.
Mathur, S., Amoeba Technologies Inc.
Originally presented on: 11/6/2013 17:15:00 - 17:35:00

One of the major challenges in the chemical industry is the optimal design of reactors for fluid mixing applications. Formation of poorly mixed zones can lead to significant losses in yield. Localization of energy can lead to losses in yield and reduced product quality, and can cause damage to the equipment. Optimal reactor design therefore requires the study of various factors, such as the geometry of the reactor, baffles and the impellers, impeller speed, flow rate of the feed, feed composition, feed location, etc., to understand their impact on the reactor performance. Traditional CFD analysis solvers typically require users to set up, solve, and view results for a single configuration. To view results when some variations are made in the original configuration, users need to set up another configuration and solve it. Comparisons can then be made based on the results obtained by solving a number of such configurations separately. This methodology can, however, become tedious when solving for numerous small variations. In addition, the traditional approach does not account for uncertainties and tolerances, both in inputs and in predictions. When the goal is to improve design and achieve a target performance, it becomes necessary for a solver to be able to account for such uncertainties and variations.

In this paper we present a numerical approach that enables computation of the sensitivity of solutions to variations in input parameters without the need for multiple single-point solutions. We use a discrete tangent methodology using operator overloading capabilities of C++ to propagate derivatives from inputs through the entire solution algorithm so that derivatives of all computed quantities with respect to any quantity of interest are determined. Thus information about the dependence on input parameters is available not only for reduced quantities such as performance metrics but for entire fields, and this can be very valuable in providing visual insight into the problem. Quantities for which sensitivity information is desired are defined in terms of a parameter and the solution as well as its derivatives are computed about the nominal value specified for this parameter. Differentiating parameters can include all the physcial parameters such as boundary conditions and material properties as well as numerical parameters such as empirical constants used in physical models and solver settings. Sensitivity analysis provides a richer information set than a single-point solution - the values and their gradients with respect to the selected parameters in the entire domain.

In this study, we will present our results for the Residence Time Distribution and flow profiles in stirred tank reactors, and their sensitivity to various parameters such as the inlet flow rate, the impeller speed, and the fluid properties. Our sensitivity results will be compared with results obtained from multiple single-point simulations using traditional solvers.

AUTHOR INDEX

Alizadeh, Ebrahim 68
Almeida, Lucilla 61
Alvarez, Mario M. 9, 35
Amedu, Osaighe Kennedy 44
Anderson, Rayford 34
Apesteguía, Carlos 23
Armenante, Piero M. 15, 28, 37
Ayranci, Inci 12
Bacchi, Raphael David A. 61
Bapat, Yogesh 70
Basavarajappa, Manjunath 45
Bashiri, Hamed 68
Benz, Gregory 36
Berson, R. Eric 71
Bertrand, Francois 38
Bertrand, François 68
Bidart, Antonio Marcos F. 61
Bigio, David 25
Blais, Bruno 38
Boesch, Steven 14
Braga, Maria 64
Brau, Christopher 51
Bredael, Gerard 37
Brunet-Errard, Léna 64
Bulnes-Abundis, David 9
Camacho-León, Sergio 35
Carrillo-Cocom, Leydi Maribel 35
Cendo, Daniel 14
Chaouki, Jamal 68
Cheng, Alan 52
Chopra, Michelle 34
Coelho, Erick F. A. Quintella 61
Cooke, Michael 17
Costa, Marcelo F. 6
De Bellefon, Claude 64
Derksen, Jos 30, 42, 46
Dias, Madalena M. 6
Dietsche, Laura J. 40
Dimattesa, Matthew 14
Dittmeyer, Roland 31
Eitzlmayr, Andreas 66
Eppinger, Thomas 62
Ershad, Navid 2
Eskin, Dmitry 42
Etchells, Arthur W. 14
Fasano, Julian B. 26
Feng, Lian-Fang 69
Fonte, Carlos M. 6
Fonte, Cláudio P. 64
Fradette, Louis 12, 18
García-López, Erika 35
Garza-García, Lucía D. 35
Ge, Chun-Yan 69
Gel, Aytekin 57
Giacomelli, Jason J. 1, 53
Grenville, Richard K. 1
Gu, Xue-Ping 69
Heniche, Mourad 68
Hesketh, Robert P. 14
Hidema, Ruri 3
Higbee, Robert 29
Higbee, Robert W. 1
Hirata, Yushi 3
Hsu, Ruey Chi 50
Hule, Pritam 70
Janz, Eric E. 26, 49
Jones, Nephi 51
Jurtz, Nico 62
Kang, Jer Chang 50
Kasat, Gopal 36, 70
Khinast, Johannes G. 32, 66
Kölbl, Andreas 31
Komoda, Yoshiyuki 3
Komrakova, Alexandra E. 42
Koot, Mathijs 24
Koscher, Gerold 66
Kraume, Matthias 62
Kraut, Manfred 31
Kresta, Suzanne 2, 27, 48
Lee, Sang Jin 29
Lin, Wenglong 34
Lopes, José Carlos B. 6, 64
Machado, Marcio Bezerra 27, 48
Mathur, Sanjay 72
Mattar, Sally 48
Meyer, Bernd 60
Miskovic, Sanja 45
Morbidelli, Massimo 65
Moreira, Rodrigo P. M. 61
Motamedvaziri, Shilan 15, 28
Mundhwa, Mayur 59
Myers, Kevin J. 26, 49
Nieves-Remacha, Maria Jose 54
Nikrityuk, P. A. 60
Nixon, Jason 25
Ornek, Dogan 34
Parmar, Nilesh 40
Passarelli, Fabio M. 61
Patterson, Gary K. 22
Peppley, Brant. A. 59
Pitault, Isabelle 64
Qayyum, Abdul 31
Radl, Stefan 66
Ramachandran, Arun 19
Ramsey, Philip 34
Regalbuto, John A. 10
Regalbuto, John R. 10
Register, Justin 10
Rodgers, Thomas L. 17
Rodríguez-González, Ciro Angel 35

AUTHOR INDEX

Ropelato, Karolline 61
Rosendall, Brigette 21
Russ, David C. 71
Santos-Moreau, Vania 64
Sathyamurthy, Prabhu 72
Sauvageau, Dominic 48
Schulze, Sebastian 60
Senda, Saki 3
Shahnam, Mehrdad 57
Shaik, Muneeb Ur Rahman 49
Simoëns, Serge 64
Soos, Miroslav 65
Srinivasan, Vibha 72
Subramaniyan, Arun K. 57
Suzuki, Hiroshi 3
Takeda, Hiroshi 3
Tantikul, Tawan 32
Tapia-Mejía, Eduardo J. 35
Thurgood, Christopher P. 59
Tsabet, Emir 18
Van Den Akker, Harry E. A. 24
Vedapuri, Damodaran 36, 70
Vonka, Michal 65
Wang, Jia-Jun 69
Wang, Li 11
Wardle, Kent E. 41, 47
Witz, Christian 32
Wu, Binxin 29, 43, 69
Wyczalkowski, Wojciech 53
Yang, Chao 13
Zhang, Yiran 37
Zhou, Ying 52